T0331703

Genetics and Philosophy

In the past century, nearly all of the biological sciences have been directly affected by discoveries and developments in genetics, a fast-evolving subject with important theoretical dimensions. In this rich and accessible book, Paul Griffiths and Karola Stotz show how the concept of the gene has evolved and diversified across the many fields that make up modern biology. By examining the molecular biology of the 'environment', they situate genetics in the developmental biology of whole organisms, and reveal how the molecular biosciences have undermined the nature/nurture distinction. Their discussion gives full weight to the revolutionary impacts of molecular biology, while rejecting 'genocentrism' and 'reductionism', and brings the topic right up to date with the philosophical implications of the most recent developments in genetics. Their book will be invaluable for those studying the philosophy of biology, genetics, and other life sciences.

PAUL GRIFFITHS is University Professorial Research Fellow at the University of Sydney. He is the author of *What Emotions Really Are: The Problem of Psychological Categories* (1997) and *Sex and Death: An Introduction to the Philosophy of Biology* (with K. Sterelny, 1999). He is the editor of *Trees of Life: Essays in Philosophy of Biology* (1992) and *Cycles of Contingency: Developmental Systems and Evolution* (with S. Oyama and R. D. Gray, 2001).

KAROLA STOTZ is an Australian Research Fellow at the University of Sydney. She has published extensively on philosophical issues in the life sciences, particularly genetics and molecular biology.

Cambridge Introductions to Philosophy and Biology

General editor
Michael Ruse, Florida State University

Titles in the series
Derek Turner, *Paleontology: A Philosophical Introduction*
R. Paul Thompson, *Agro-technology: A Philosophical Introduction*
Michael Ruse, *The Philosophy of Human Evolution*
Paul Griffiths and Karola Stotz, *Genetics and Philosophy: An Introduction*

Genetics and Philosophy

An Introduction

PAUL GRIFFITHS
University of Sydney

KAROLA STOTZ
University of Sydney

CAMBRIDGE
UNIVERSITY PRESS

University Printing House, Cambridge CB2 8BS, United Kingdom

One Liberty Plaza, 20th Floor, New York, NY 10006, USA

477 Williamstown Road, Port Melbourne, VIC 3207, Australia

314-321, 3rd Floor, Plot 3, Splendor Forum, Jasola District Centre, New Delhi - 110025, India

103 Penang Road, #05-06/07, Visioncrest Commercial, Singapore 238467

Cambridge University Press is part of the University of Cambridge.

It furthers the University's mission by disseminating knowledge in the pursuit of education, learning and research at the highest international levels of excellence.

www.cambridge.org
Information on this title: www.cambridge.org/9781107002128

First published 2013
Reprinted 2014

A catalogue record for this publication is available from the British Library

Library of Congress Cataloging in Publication data
Griffiths, Paul, 1962–
Genetics and philosophy : an introduction / Paul Griffiths, University of Sydney, Karola Stotz, University of Sydney.
    pages   cm. – (Cambridge introductions to philosophy and biology)
Includes bibliographical references and index.
ISBN 978-1-107-00212-8 (hardback) – ISBN 978-0-521-17390-2 (paperback)
1. Genes.   2. Genomics.   3. Genetics – Philosophy.   4. Developmental genetics.
I. Stotz, Karola.   II. Title.
QH447.G75   2013
572.8´6 – dc23   2012042715

ISBN 978-1-107-00212-8 Hardback
ISBN 978-0-521-17390-2 Paperback

Contents

Figures and tables

Figures

Tables

Acknowledgments

We would like to thank the members of the Philosophy and History of Biology group at the Sydney Centre for the Foundations of Science for feedback on early drafts. Staffan Müller-Wille gave us valuable comments on Chapters 2 and 3, Arnon Levy on Chapter 6, and James Tabery on Chapter 7, which in any case draws extensively on earlier collaborative research by Tabery and Griffiths. Peter Godfrey-Smith provided helpful comments to the final manuscript, as did Isobel Ronai, who also prepared the index.

The research for this book was supported under the Australian Research Council's Discovery Projects funding scheme, project number DP0878650.

Cover image: The ability of a DNA molecule to act as a collection of genes depends on many other molecules. This picture shows DNA being inactivated (switched off). The DNA double helix (purple) is in the process of being coiled tightly around histone molecules (blue-white) to form 'nucleosomes'. Once this is complete the DNA cannot be used as genes until the process is reversed. Thin 'tails' project from each cluster of histone molecules. Chemical modifications to these tails (bright yellow and turquoise) are produced by interactions with other molecules in the cell. It is these changes that control the processes of activation and inactivation. *Image courtesy of Etsuko Uno and the Walter and Eliza Hall Institute of Medical Research, Melbourne, Australia.*

1 Introduction

Unlike most books which combine philosophy and genetics in their titles, this is not a discussion of the ethical, legal, and social implications of science. It is a contribution to the philosophy of science, the branch of epistemology (theory of knowledge) which sets out to understand how science works. The word 'genetics' is construed broadly to include a wide range of molecular biosciences, and the exposition of these sciences is a backdrop to our discussion of the philosophical issues of reductionism and reductive explanation, the status of theoretical entities, and the relationship between scientific representations – models – and the targets of those representations. Genetics and molecular biology have been a powerful source of philosophical insights into these issues. Recent scientific developments in this rapidly changing area hold new lessons for philosophy of biological science.

Since Aristotle philosophers and scientists have reflected on the nature of living systems and the distinctive nature of the sciences that study them. However, the emergence of the philosophy of science as a distinct academic field in the early twentieth century was marked by an almost exclusive focus on the physical sciences. When philosophers of science turned their attention to biology in the 1960s, one of the first issues to be raised was whether the new molecular biology constituted a successful reduction of earlier biological theories, and particularly earlier theories of genetics (Schaffner 1967; Schaffner 1969; Ruse 1971; Hull 1972; Hull 1974). As well as addressing general issues like reduction, philosophers of science are tasked with analysing key scientific concepts, and the concept of the gene has proved both attractive and elusive. In part this is because it is a moving target. The concept of the gene had evolved considerably in the years between the introduction of the concept at the turn of the twentieth century and the papers just cited, and it has continued to evolve during the past forty years of intense philosophical attention.

Many of the classic philosophical papers on reductionism in molecular biology date from the 1980s, or continue the debate in the terms established in that period (Kitcher 1982; Kitcher 1984; Rosenberg 1985; Wimsatt 1986a; Waters 1990; Schaffner 1993; Waters 1994). The molecular conception of the gene which figured in these debates was the temporary consensus around the 'classical molecular gene' concept, which we describe in Chapter 3. However, the 1990s and 2000s saw the discovery of far greater complexity both in how genes are structurally constituted in the genome, and in how genes function to make their products. In the 'postgenomic era', when complete genome sequences are available for an increasing range of organisms, the range of molecular actors has expanded greatly. The genome is not merely a collection of genes, but houses diverse other functional elements. Genes no longer have a single function closely related to their structure, but respond in a flexible manner to signals from a massive regulatory architecture that is, increasingly, the real focus of research in 'genetics'. One of the main aims of this book is to revisit those earlier philosophical debates against this very different scientific background.

This is not a history of genetics. But science is a dynamic process, and understanding it often involves understanding how concepts and theories have changed. So in some places we do give historical treatments of the emergence, development, and sometimes abandonment of ideas in genetics. In such cases, we have drawn heavily, and we hope with sufficient acknowledgment, on the many historians of science who have devoted themselves to genetics and molecular biology.

We have chosen not to deal with population genetics, the discipline whose primary focus is the algebraic consequences of Mendelian heredity and selection in populations. There are two reasons why this would take us into a very different philosophical territory from the book we have written. The first is that population genetics is a mathematical discipline centred on a few principles of high generality: the structure of population genetic theory has frequently been compared to the structure of theories in physics (Sober 1984; Brandon and McShea 2010). This is not a coincidence, since one of the creators of population genetics, Ronald Aylmer Fisher, modelled his theory on statistical thermodynamics (Depew and Weber 1995). Philosophical analyses of the molecular biosciences, in contrast, have shown that these sciences do not have a mathematical, or plausibly mathematicisable, core of highly general claims (Darden and Maull 1977; Bechtel and Richardson 1993; Schaffner

1993; Schaffner 1996; Bechtel 2006). Instead, they are organised around a cast of evolutionarily conserved parts and processes, in a way more comparable to sciences such as physiology and anatomy (Winther 2006). A second reason why a discussion of population genetics would have taken us too far afield is its close relationship to evolutionary theory. Population genetics is the mathematical core of modern evolutionary biology. A philosophical discussion of population genetics would have to engage with the philosophy of evolutionary biology, still the largest area in the philosophy of biology. This is the topic of another volume in this series, and more than deserves a book to itself. However, in Chapter 8 we do discuss some of the ways in which the developments in molecular biology outlined in this book are likely to produce, as they are more fully assimilated, changes in our understanding of evolution and how those changes may be reflected in evolutionary theory.

In Chapter 2, 'Mendel's gene', we begin our exploration with an account of the emergence of genetics at the beginning of the twentieth century. In line with much recent scholarship we argue that the important element of Mendelian genetics was not a few 'laws' of heredity, but the experimental practice known as 'genetic analysis'. We show how genetic analysis was used to solve problems in many other areas of biology. Following the historian Raphael Falk, we argue that from its very introduction the gene had two identities. The first, and initially the most prominent, was that of an instrumental unit defined by its role in genetic analysis. The second identity was that of a hypothetical material unit of heredity (Falk 1984, 1986, 2009). In their instrumental identity the existence of Mendelian genes is guaranteed by the success of genetic analysis. Hence Mendelian genes were never merely hypotheses whose confirmation awaited the discovery of the material gene. We compare the ontological status of the Mendelian gene in its instrumental identity to that of centres of mass in dynamics. Building on this approach, we argue that geneticists in the first decades of the twentieth century had two ways of thinking about – representations of – the gene. Thinking about the gene as an instrumental entity was useful in the context of genetic analysis. Thinking about the gene as a hypothetical material entity was increasingly useful as geneticists came closer to understanding the material basis of heredity.

In Chapter 3, 'The material gene', we describe how the elucidation of the structure and basic function of DNA represented the successful conclusion of the search for the gene as a material unit of heredity: the way in which DNA is passed from one cell to the next provides the physical underpinnings for the gene's instrumental role as marker of phenotypic differences across generations. However, the causal role of the gene as it had been envisaged in classical genetics was very substantially revised in order to fit what had been discovered about the material basis of heredity. The result of the molecular revolution in genetics was not that a causal role (the Mendelian gene) was filled by a material occupant (the molecular gene). The molecular gene had a new role, very different from that of the Mendelian gene. Its primary role was to specify the linear order of elements in cellular products, initially polypeptide chains, the precursors of proteins. This explains the difficulties encountered by philosophers who have tried to explain how the Mendelian gene was reduced to molecular biology. Although the new, molecular identity of the gene was now its dominant identity, the other, instrumental identity did not simply go away. The original role of the Mendelian gene continues to define the gene in certain areas of biological research: namely, those intellectually continuous with classical genetic analysis. We give examples of contemporary research in which it is necessary to think of genes as both Mendelian alleles and molecular genes, even when those two identities do not converge on the same pieces of DNA. Reductionists are correct that the gene turned out to be grounded in DNA, but they fail to recognise that the development of genetics has left us with more than one scientifically productive way of thinking about DNA and the genes it contains. This is in large part because they have failed to recognise how the different identities of the gene are anchored in different experimental practices.

The other major theme of Chapter 3 is the emergence of 'informational specificity' as the key property of the molecular gene. We describe how biological specificity (the ability of biomolecules to catalyse very specific chemical reactions) was transformed from a physical concept based on stereochemistry (the three-dimensional shape of molecules) to an informational concept based on the linear correspondence between molecules, most famously in the case of the genetic code. We introduce the term 'Crick information' to refer to the sense of 'information' introduced by Francis Crick (1958) and used to define informational specificity.

Chapter 3 introduces a philosophical model of explanation which will recur throughout the remainder of the book as the best way to capture the nature of research in the molecular biosciences. Following recent neo-mechanist philosophers we argue that mechanistic explanation includes both a reductionist phase and an integrative phase. The reductionist phase of research identifies and characterises the constituent parts of a mechanism. The integrative phase shows how the phenomenon to be explained is produced by the specific ways in which those parts are organised so as to make up that mechanism (Bechtel and Abrahamsen 2005; Bechtel 2006; Craver and Bechtel 2007).

In Chapter 4, 'The reactive genome', we explore one dimension of the increased complexity of 'postgenomic' biology. We argue that informational specificity or Crick information – the ability to causally specify the linear sequence of a gene product – is not located solely in coding sequences of DNA, but is distributed between the coding sequences, regulatory sequences and their RNA and protein products, and the environmental signals that act via that regulatory machinery. These other factors help to determine the specificity of gene products through the activation and selection of coding sequences, and the creation of additional Crick information during post-transcriptional processing. We outline the concepts of 'distributed specificity' and 'combinatorial control' and show that they support a profoundly non-reductionist account of gene function which we refer to as 'molecular epigenesis' (Burian 2004; Stotz 2006a). The way in which genes in combination with other actors determine the activity of cells is mechanistic, but it is not reductionistic.

In Chapter 5, 'Outside the genome', we look at the sources of the environmental signals which act as drivers for genome expression and are an additional source of Crick information, and explore the new fields of 'epigenetics' and 'epigenetic inheritance' in both the narrow and wider senses of those contested terms. Genetics as the study of heredity has traditionally been aligned with the nature side of the nature/nurture dichotomy, which has in turn been regarded as 'reductionist', while scientists who have focused on nurture have been labelled as 'anti-reductionist'. Today, however, some aspects of nurture have proved to be heritable, and in addition the study of nurture has gone increasingly molecular, so that research into the role of the environment in the development and functioning of organisms is potentially as 'reductionist' – that is to say, mechanistic – as research in any other areas of the molecular biosciences. Organisms construct their life cycles through the interaction of the contents of the fertilised egg, the genome and its narrowly epigenetic surroundings, with a 'developmental niche' which is the result of epigenetic inheritance in a wider sense (to avoid confusion, we refer to this as 'exogenetic inheritance'; West and King 1987). Organisms inherit elements of their developmental niche in much the same sense that they inherit their genome, albeit via different mechanisms of transmission. We reiterate our argument that the nature of development supports mechanistic anti-reductionism: developmental outcomes are explained by the organisation of the components which regulate gene expression, but cannot be reduced to the components taken out of the context of their causally crucial organisation. The regulatory architecture of the genome extends outside the organism into the developmental niche, partly vindicating some other traditional 'anti-reductionist' themes.

In Chapter 6, 'The informational gene', we discuss genetic information, the genetic programme, and the informational identity of the gene. This is the conception of genes as units of information, supplying the form to complement matter and make matter come to life. The informational identity of the gene provides the underlying rationale for the view that genes retain a unique importance in development despite all evidence of the impact of other factors. Attitudes to the idea that biology is an 'information science' differ profoundly: some regard it as the greatest insight of twentieth-century biology, others as no more than a muddle caused by taking metaphors too seriously. Our position is somewhere in between. We argue strongly against *semantic* conceptions of both genetic information and the genetic programme – those which seek to identify meanings and messages in molecules. However, we conclude that in reacting against these semantic approaches, critics such as ourselves have mistakenly dismissed less overblown but very important informational ideas in biology. While we and other critics have insisted on restricting talk of the 'genetic code' to the actual triplet code which translates nucleic acid into protein (Godfrey-Smith 2000a; Griffiths 2001), there is more to say about this than we had supposed. We now propose that the code is a means to transfer information in the sense defined by Francis Crick (Crick 1958, 1970): namely, sequence specificity, or Crick information. Crick information is not contained solely in nucleic acid sequence, as the previous chapters will establish. However, despite the existence of other mechanisms of inheritance the ability of organisms to transfer sequence specificity between generations is crucially dependent on the invention of nucleic acid-based heredity. Nucleic acid-based heredity is an evolutionary 'key innovation' because it allows secure and efficient transfer of specificity between cells (similar emphases exist in the accounts of Moss 2003 and Sarkar 2005). We also take on board the criticism that philosophers like ourselves have not appreciated the theoretical value of treating heredity as a formal coding problem using the mathematical theory of communication (Bergstrom and Rosvall 2009). We argue that the information whose transmission is being optimised here is, once again, Crick information. Finally, we argue that the concept of a genetic programme can and should be divorced from the traditional idea of the genome as a 'blueprint' for the organism (Mayr 1961). The genetic programme as it figures in contemporary molecular developmental biology is best understood as a form of mechanistic explanation corresponding to the concepts of distributed specificity and combinatorial control described in Chapter 4.

In Chapter 7, 'The behavioural gene', we look at the use of genetics to explain behaviour, including human behaviour. This chapter draws on earlier, collaborative research with James Tabery. We briefly revisit the well-trodden ground of the interpretation of heritability coefficients and the other results of traditional statistical behaviour genetics. Our primary concern, however, is to explain how genes and gene action were conceptualised by traditional, quantitative behaviour geneticists and by their critics from the science of behavioural development. Behaviour geneticists and their critics focus on two different identities of the gene. Whereas behaviour geneticists use a Mendelian representation of the gene, their critics think in terms of what we call 'abstract developmental genes'. These two representations of the gene feature in two very different styles of genetic explanation of phenotypes. We show how the substantial scientific disagreements between these two groups were grounded in these differences. Chapter 7 also shows how the integration of molecular methods into both behaviour genetics and its traditional adversary developmental psychobiology has created common ground on which their differences can be resolved through research rather than polemic.

Finally, in Chapter 8, 'The genome in evolution', we discuss how evolutionary theory may be affected by the research discussed in the preceding chapters. The message here is that some of the assumptions underlying the 'Modern Synthesis' are based on an outmoded conception of the genome, and are significantly challenged by new developments in the molecular biosciences.

We have no illusions that this book will be the last contribution to the forty-year philosophical discussion of the gene, but we do believe that our conclusions move that discussion forward. Briefly, we argue that the gene today has several identities, identities which have accumulated as the molecular biosciences have developed and diversified. It is still an instrumental unit for genetic analysis, and it is also a reasonably clearly defined structural unit used in annotating genomes. The gene is also a unit of Crick information, but the relationship between this identity of the gene and its conventional structural definition has become increasingly vexed in recent years. It also has less prominent identities: in Chapter 7 we show that some 'genes' are no more than hypothesized anchors for the parameters of developmental models. Each of these identities plays a productive role in some forms of biological research. Scientists are adept at thinking about genes in whichever way best suits their work, and at switching between these different representations of the gene as the nature of their work changes. The concept of the gene is therefore best conceived as a set of contextually activated representations.

Our other conclusion is that recent developments in the molecular biosciences have considerably undermined the idea that genes, however understood, are the prime movers in all biological processes. Despite the key role of nucleic acid inheritance in making it possible to move biological specificity between the generations, there is much more to heredity than the inheritance of nuclear DNA. Although all biomolecules are ultimately synthesised from a nucleic acid template, that template is only one source of the specificity of those biomolecules. Finally, despite the importance of gene control networks in the regulatory architecture of the cell, the complete regulatory apparatus includes a much wider 'developmental niche'. The specific roles played by the gene in its several identities are more than enough to explain its central place in biology. There is no need for anything more grandiose.

2   Mendel's gene

2.1 The birth of the gene

Around 1900 a number of scientists observed 'Mendelian ratios' in plant and animal breeding experiments. Mendelian ratios can be observed when two varieties of a plant or animal, each of which reliably displays some observable characteristic, such as the height of a plant or the colour of its flowers, are crossed to produce hybrid offspring. In the first generation, one of the two parental characteristics disappears, and all the offspring show the other characteristic. All plants in the first generation may be tall, even if only one parent plant was tall and the other was short. Or all the offspring may have red flowers, even if one parent had red flowers and the other white. If this happens it will appear that only one of the two parental characters has been passed to the next generation, and the other character has been lost. But if these first-generation offspring are crossed with one another, the second generation will display both the characters seen in the two original varieties, and will display them in the Mendelian ratio of 3:1. Three-quarters of the second generation show the character which was universal in the first generation, while one quarter show the character which disappeared in that generation. There is a compelling explanation of this and other, more complex Mendelian ratios which hypothesises that each organism contains two factors that determine which character it will display. One factor comes from each parent, and if an organism inherits two different factors, one is always expressed preferentially over the other (Figure 2.1).

Today we are all familiar with these factors, which we know as 'genes'. But in this chapter we aim to take the reader back to the birth of the gene, and to show how this idea grew out of a particular kind of experimentation. This will have two philosophical payoffs. The immediate one is insight into the status of a certain kind of theoretical entity in science. The status of the gene in

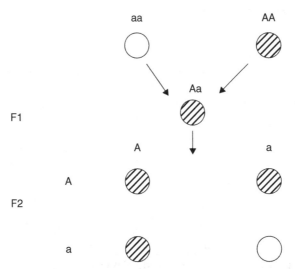

Figure 2.1 Mendelian ratios. Individuals from two inbred lines in which all individuals are dark or light, respectively, are crossed to produce a hybrid line. In the first generation (Filial 1 or F1), all the offspring are dark. When these offspring are crossed with one another the second generation (F2) contains both dark and light offspring in a 3:1 ratio, reflecting the four possible combinations of gametes from parents which are both Aa.

the early decades of Mendelian genetics is a useful corrective to simple ideas about 'unobservable' or 'theoretical' entities in science. The second payoff will become evident later in the book, as we explore other identities that the gene has acquired. We will see how the original identity of the gene from its first, strictly Mendelian context lives on alongside these other identities as one element of the complex identity of the gene today.

Mendelian ratios had already been observed and explained forty years earlier by Gregor Mendel, a scientist and Catholic monk in what is now the Czech Republic. The results of Mendel's experiments with peas were published in a respectable local scientific journal (Mendel 1866) and were reasonably well known among scientists with similar interests. But for scientists like the Englishman William Bateson and the Danish botanist Wilhelm Johannsen forty years later Mendelian ratios had a much broader significance: they revealed the basic principles of heredity. The hereditary contribution to the observable characters of an organism consists of pairs of factors, one of which comes from each parent. The two factors an organism receives from its parents remain within it, and are passed on unchanged to the next generation. Parents pass on one factor to each of their offspring, so that the offspring receive one factor from each parent.

The decades separating Mendel's own work from the rise of Mendelian genetics were a period of intense interest in the laws and mechanisms of heredity. This was partly a response to Charles Darwin's theory of evolution by natural selection. If new species arise from old ones, the principles of heredity must be such as to allow this. Moreover, an understanding of heredity should lead to a better understanding of the process of evolution. Mendel's findings received more attention the second time around because they provided answers to these new questions. Mendelians, as they now called themselves, were keen to distinguish their approach to heredity from other research traditions that had been prominent in the late nineteenth century. They wanted to distance Mendelism from what they saw as merely speculative accounts of the physical mechanisms of heredity offered by authors including Darwin himself and, more importantly, the German biologist August Weismann. They also rejected the 'law of ancestral heredity', a view of heredity defended by the leading biological statisticians of the period, who worked in the field then known as 'biometrics' which was to develop into quantitative genetics (see Chapter 7).

Biometricians thought that the aim of a science of heredity was to discover the statistical relationship between the features of an organism and the features of its offspring. They approached this problem, not through small-scale controlled experiments like Mendel's, but through the statistical analysis of large sets of data from natural populations, such as those available for human beings and for domesticated animals such as horses. While there is a correlation between the characters of parents and offspring in such data, it is far from perfect. Tall people are likely to have tall children, but not all their children need be tall and some short people have surprisingly tall children. If data about grandparents and great-grandparents are added to the data about the parents it is possible to predict the character of the offspring more accurately. This led Darwin's younger cousin, Francis Galton, to propose the 'law of ancestral heredity'. It states that one half of the character of an individual is explained by the characters of his or her parents, one quarter by those of the grandparents, and so forth. To predict a person's height, Galton would gather data on the height of his or her parents, grandparents, great-grandparents, and so on, and calculate the average height of the two parents, the average height of the four grandparents, of the eight great-grandparents, and so forth.

He would weight the average heights in each generation by half for the parents, quarter for the grandparents, and so on, and add the results to get a prediction for the person's height. The same procedure can be applied to predict how far an individual will differ from the population average, using data on how far their ancestors departed from the average in their own generation, weighted in the same way. To apply this procedure fully would require data on an infinite number of generations, so that the series of weights – half, quarter, eighth, sixteenth, and so on – would total 1. In practice, only data on the first few generations are available. This does not matter because the other generations make only a tiny contribution: data from ten generations back are already weighted at only 1/1024.

Galton's more mathematically accomplished follower, Karl Pearson, argued that the law of ancestral heredity was not a biological law at all, but an application of statistics which could be used to predict any variable from a set of variables with which it is correlated. Consequently, Pearson argued, the weights of half, quarter, and so on which Galton had assigned to each generation of ancestors were only estimates. The real weights would have to be determined empirically from large data sets, and they might differ between species, or be different for different kinds of characters.

Pearson's ideas about scientific method were strongly influenced by the 'instrumentalist' views of the Austrian physicist Ernst Mach. Science is a way of systematising observations, and any attempt to describe unobservable entities which underlie the observable phenomena strays from science into metaphysics. The scientific method can be summarised as 'the orderly classification of facts followed by the recognition of their relationship and recurring sequences' (Pearson 1900, 18–19). Scientists may use theoretical terms ('conceptions', as Pearson called them), but they must never suppose that these 'conceptions' correspond to unobservable entities. Mach had criticised the atomic theory of matter for straying into metaphysics, and Pearson criticised his predecessors in the theory of heredity on the same grounds. Theories about particles transmitted from parent to offspring can be scientific only if the 'particles' are no more than a way of describing the observable resemblance between parent and offspring. Unfortunately, said Pearson, 'in the theories of both Darwin and Weismann a metaphysical element seems to enter'. This is because they treat the particles, the unobservable mechanisms of heredity, as the actual objects of their research and not merely as an instrument to handle the observable data (Pearson 1900, 337). In light of these views, it is not surprising that Pearson was unreceptive to Mendelism and its equally unobservable 'factors'.

From the Mendelian point of view Pearson and other biometricians were looking at the whole subject of heredity in the wrong way. Their focus on the observable characters of parent and offspring was fundamentally misguided. A yellow-flowered plant does not literally inherit yellow flowers from its parents. The plant develops from a seed, and whatever it inherits must be in that seed. But the seed does not contain the colour of the flowers. Instead, the seed contains something which influences the later growth of the plant, including the colour of its flowers. Johannsen (1911) introduced some new terminology to make this clear. What an organism inherits from its parents is only its 'genotype', and not its 'phenotype'. A plant's genotype is entirely determined by the male pollen and female ovule which merge to form the seed. The genotype of an animal depends on nothing but the egg and the sperm which fertilised it. The phenotype of the plant or animal, however, is a character that develops from the seed or egg at a later point in life – such as the colour of the flowers, or the height of the plant. Many other factors influence the growth of the plant. If these factors are similar for parents and the offspring, then the offspring will achieve a similar height to that of the parents. If those other factors are different – for example, if the parents grew in a wet year and the offspring in a dry year – then the offspring may be much shorter than the parents. But this fact is irrelevant to what the plant inherited from its parents – its genotype. The relationship between parent and offspring phenotypes, Johannsen argued, is not the real phenomenon of heredity. It merely provides evidence that we can use to investigate the relationship between the parent's genotype and the offspring's genotype.

While this is now the conventional way to think about heredity, it was revolutionary when Johannsen introduced it. In Figure 2.1 it can be seen that a characteristic disappears in the first generation and reappears in the next. Writing a few years before Johannsen introduced the distinction between genotype and phenotype, Pearson used this phenomenon to argue against Mendelism. According to Mendelism, heredity is restricted to what parents pass on to their offspring in the seed or in the egg. But simple inspection of the data shows that offspring are affected by more than one generation of their ancestors, so Mendelism must be mistaken (Pearson et al. 1903). Pearson was no fool, but when he made this argument he thought of heredity as a statistical relationship between phenotypes, whereas the Mendelians were coming to think of it as a deterministic relationship between genotypes. The genotype of an organism is determined solely by the genotype of the two immediate parents.

Johannsen introduced the term 'gene' to refer to the things which make up an organism's genotype. Each gene comes in a number of alternative forms, known as 'alleles'. An organism contains a number of loci (places), one for each gene. Each locus can hold two alleles of that gene. An organism's genotype at a single locus is this pair of alleles. These may be two copies of the same allele (the organism is 'homozygous' AA or aa), or one copy of each of two different alleles ('heterozygous' Aa or aA). An organism's overall genotype is the combination of all the pairs of alleles at all the loci. In the very simplest Mendelian models, like those constructed by Mendel himself, there are only two alleles of each gene. Mendel's 'law of segregation' says that each parent passes on only one allele to each offspring. Mendel's 'law of independent assortment' says that which allele an organism gets from one locus in the parent has no effect on which allele it gets from another locus. In other words, how alleles segregate at one locus is independent of how they segregate at another locus. Mendel also assumed that for every pair of alleles, one allele is 'dominant' and the other 'recessive', meaning that if an organism has one copy of each allele, it will look like organisms that have two copies of the dominant allele. When Mendel's two laws hold, and one allele at each locus is dominant, then a Mendelian model of heredity for two different loci immediately predicts the famous 9:3:3:1 ratio that Mendel observed in the F_2 generation in his peas. For every nine offspring with both the dominant phenotypic characters, there will be three which have one recessive phenotypic character, three which have the other recessive phenotypic character, and one which has both recessive phenotypes.

2.2 The Mendelian gene

As well as rejecting the law of ancestral heredity, the Mendelians distanced themselves from earlier proposals about the physical mechanisms by which the contents of the gametes (the pollen and ovule in plants, or the sperm and egg in animals) determine those of the zygote (the pollinated ovule or the fertilised egg), and the mechanisms by which the contents of the zygote affect the characters of the organism which develops from that zygote. In particular, they wanted to disassociate themselves from theories developed by Weismann in the 1880s about material particles which pass from one cell to the next through the formation of the sex cells and their fusion to create a zygote, and which go on to produce the many different types of cell needed to make up a whole organism. There was no direct evidence for these particles and their postulated behaviour, whereas the Mendelians insisted that every aspect of their theory could be backed up with experimental evidence: 'Mendelians have the great merit of being prudent in their speculations [...] a quite natural reaction against the morphologico-phantasmagorical speculations of the Weismann school' (Johannsen 1911, 133).

The entity at the heart of Mendelian genetics – the gene – had a very distinctive status. It was not an observable entity, but it was something more than an unobservable entity postulated to explain the data. The gene was a *tool* for predicting the outcome of breeding one organism with another. Mendelian genetics cannot be made to work while referring only to the observable phenotypic traits and the relationships between them. The relationships between phenotypes are mediated by relationships between genotypes. If the letters representing the alleles in each genotype were removed from Figure 2.1, leaving only the pictures of the phenotype, then it would be impossible to fill in the genotypes of the next generation, and hence impossible to calculate their expected phenotypes. So the gene was not merely postulated to explain why Mendelian genetics worked, it was an essential tool for making it work.

It was natural to hope that the gene would one day be shown to exist as a physical reality within the cells of the organism, and many Mendelians were firmly committed to this idea. But the special role of the gene in the practice of Mendelian genetics meant that the gene would remain an important and legitimate idea even if this did not work out. Thomas Hunt Morgan received the Nobel Prize in 1933 for his work demonstrating that genes are distributed along chromosomes. Nevertheless, in his Nobel lecture he noted: 'There is no consensus of opinion amongst geneticists as to what the genes are – whether they are real or purely fictitious – because at the level at which the genetic experiments lie, it does not make the slightest difference whether the gene is a hypothetical unit, or whether the gene is a material particle' (Morgan 1934). The possibility that Morgan was so careful to leave open was that genes might turn out to be something like centres of mass in physics. When two bodies act on one another – for example, by being at the two ends of a lever – their masses are distributed throughout each body. But to calculate how the bodies will affect one another, we pretend that the whole mass is located at a single, infinitesimal point – the centre of mass of that body. The centre of mass is not a material particle. If you stick something to one side of a body, its centre of mass moves, but nothing actually changes at either the old or the new location of the centre of mass. Nevertheless, every object really does have a centre of mass. The only way to show that centres of mass do not exist would be to refute principles of mechanics that have held sway since Archimedes, something that is, to say the least, unlikely to happen.

Similarly, to show that Mendelian genes do not exist it would be necessary to show that the results obtained in countless breeding experiments were mistaken, or that better results could be obtained using an alternative theory which does not involve pairs of factors which segregate, are dominant to one another, and so forth. So although the early Mendelians produced models in which hypothetical particles move from one cell to another, they were engaged in a very different kind of science from earlier biologists who had theorised about such particles. Those earlier scientists had postulated the existence of particles to explain existing observations about heredity and development. Darwin called his hypothesised particles 'pangenes'. Some experiments by Galton soon suggested that Darwin's pangenes were not actual, physical objects. Because the only role of Darwin's pangenes was to explain pre-existing observations, these experiments suggested that his explanation of heredity in terms of pangenes was simply wrong. The Mendelians were confident that the existence of genes could not be refuted in such a simple way. Even if there were no straightforward physical particles corresponding to genes, genes would still be essential devices for calculation.

The historian of genetics Raphael Falk has summed up the situation by saying that the gene of Mendelian genetics – or 'classical genetics' as it is often called in its mature form – had two separate identities (Falk 1984, 2009). One identity was as a hypothetical material entity, and some genetic research was directed to confirming the existence of these entities and finding out more about them. But the gene had a second, and more important, identity as an instrumental entity – a tool used to do biology. If this had been the only identity of the gene in the classical, Mendelian tradition then genetics would have satisfied all the strictures of Pearson's instrumentalist account of science. But this was only one of two identities of the gene, and the future development of genetics was the result of the interplay between the two.

Mach, Pearson, and their more sophisticated successors, the logical positivist philosophers of the inter-war years, and the logical empiricists of the

1950s and 60s, took the primary unit of scientific achievement to be a well-confirmed theory. Science proceeds by gathering evidence until a theory is solidly confirmed. Thereafter, the theory can be applied, either for scientific purposes or for practical, technological purposes. The title of Morgan's *The Theory of the Gene* (1926) might suggest that the first quarter-century of Mendelism was spent confirming the theory, and the second quarter applying it. But looking at Mendelian genetics as a theory comparable to Newton's theory of gravity or Maxwell's theory of electromagnetism sells it very short. The primary contents of the theory are Mendel's two 'laws' of segregation and independent assortment, both of which are only applicable to some organisms, and the chromosome theory of the physical basis of inheritance which explains some of the key exceptions to the two laws. Much of Morgan's book is devoted to the effect of duplications of chromosomes, which can produce apparent exceptions to basic Mendelian principles. Moreover, the Mendelian approach was soon applied to asexual organisms, in which segregation and assortment do not occur, although geneticists found ingenious ways to produce analogues of those processes for experimental purposes, as we will see below. This suggests another way to look at Mendelian genetics. Most recent scholars agree that the real achievement of classical Mendelian genetics was not a theory centred on a few principles of high generality, but rather an experimental tradition in which the practice of hybridising organisms and making inferences from patterns of inheritance was used to investigate a wide range of biological questions. Geneticists investigated many aspects of an organism's biology by subjecting them to 'genetic analysis'. This means working out how the relevant characters of the organism are inherited – how many loci are needed, how many alleles exist for each locus, and how those alleles interact. The aim of this work was not to test the Mendelian theory, or to simply apply that theory, since many of the most interesting discoveries were made when the theory failed to work as expected. The aim was to find out more about the biology of the characters being analysed.

An important early example of genetic analysis was Morgan's analysis of the white-eye mutant in the fruit fly *Drosophila melanogaster*. The first Mendelians discovered almost at once that Mendel's law of independent assortment was not accurate. In many cases, which allele a parent contributes to a gamete at one locus is correlated with which allele it contributes at another locus. These loci are said to be 'linked'. Linkage would eventually be explained as the result of genes being physically linked together on chromosomes.

Independent assortment only holds for loci which are on different chromosomes. But the relationship between chromosomes, physical features which had been observed in the cell nucleus, and genetics first had to be established. An important step in demonstrating that relationship was Morgan's genetic analysis of the white-eye mutant.

Crossing a mutant, male white-eyed fly with a normal ('wild-type') red-eyed female fly produced an F1 generation consisting entirely of red-eyed flies, as would be expected, since most mutations are recessive. Basic Mendelian principles of the kind described above predict that a cross between these F1 flies will produce the dominant red-eyed character and the recessive white-eyed character in the Mendelian ratio of 3:1 (Figure 2.1). It did, but all the flies with white eyes were male, a striking fact that calls out for some explanation. Morgan also crossed a heterozygous red-eyed female with a white-eyed male. Mendelian principles imply that this female carries one unexpressed white-eye allele and one red-eyed allele, while the male carries only white-eyed alleles. This cross produced male red-eyed, female red-eyed, male white-eyed, and female white-eyed flies in equal numbers: that is, in the ratio 1:1:1:1. This demonstrated that it is possible for females as well as males to show the white-eye mutant phenotype. Females are white-eyed if they receive two white-eye alleles, one from each parent, just as basic Mendelian principles would suggest. So why were there no white-eyed females in the F1 generation in the previous experiment?

The red-eyed males used in the cross which produced the unexpected result were the offspring of a wild-type, red-eyed female and a white-eyed male. Since the white-eye allele is recessive, the basic principles of Mendelism suggest that this white-eyed male had two copies of the white-eye allele and passed on one to each of its offspring. So each of the red-eyed males used in the second cross should have carried a recessive, white-eyed allele. This allele should have been passed on to half of their female offspring, and half of those females should have received a second copy of the white-eyed allele from their heterozygous mother and been white-eyed. But this did not happen – there were no white-eyed females.

The result would make sense if the red-eyed males used in the second cross did not, in fact, have a recessive white-eyed allele. A potential way in which this could occur had already been suggested when Morgan carried out these experiments. The suggestion derived from microscopic observations of chromosomes. Chromosomes usually come in identical pairs, but not always. In female *Drosophila*, all the chromosomes come in identical pairs, but male flies have one pair of non-identical chromosomes. It was suggested that the aberrant member of this chromosome pair, not seen in female flies, was responsible for the difference between male and female. This is the now familiar idea that many organisms, including humans and *Drosophila*, have a chromosomal sex-determination system in which females have two X chromosomes and males have one X and one Y chromosome. This idea, however, had yet to be generally accepted. Morgan and others were not even convinced at the time that genes were carried on the chromosomes. The evidence to support the idea that genes are carried on chromosomes was primarily circumstantial – chromosomes, like alleles, come in pairs and one member of each pair goes to each gamete. However, if genes are, indeed, carried on chromosomes, and if the white-eye locus is located on the X chromosome, then males would only have one allele at this locus because they only have one copy of the X chromosome. This would explain the results of Morgan's experiments.

To test this explanation, Morgan carried out a series of further crosses. In one of these he crossed a white-eyed (homozygous recessive) female with a red-eyed male (Figure 2.2). The male offspring of this cross would have only one X chromosome, which it would receive from the mother. So if genes are carried on chromosomes, and the white-eye locus is on the X chromosome, then the male offspring would all be white-eyed. Conversely, female offspring of this cross would all receive the single X chromosome from the red-eyed father, carrying the dominant red-eyed allele, and would have red eyes. The results of this and Morgan's other tests were just as predicted.

So by analysing the genetic structure of the white-eyed mutant in *Drosophila*, Morgan not only established the phenomenon of sex-linked inheritance, but also provided powerful evidence for the chromosome theory of inheritance. This was the point of the series of experiments. They were not designed to test the basic principles of Mendelian heredity, such as the idea that there are discrete genetic loci which can be occupied by one or more alleles of the same gene. Morgan's experiments exemplify the idea that genetic analysis was a tool of biological enquiry. Classical genetics was not a theory under test, or a theory that was simply applied to produce predictable results. It was a method of expanding biological knowledge.

Even before full acceptance of the chromosome theory, breeding experiments had revealed the existence of 'linkage groups' of genetic loci. Mendel's law of independent assortment holds for loci in different linkage groups.

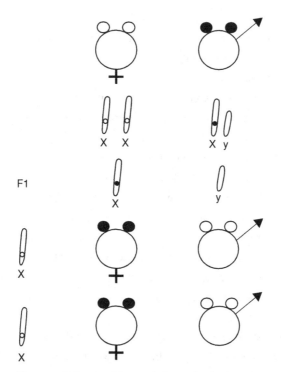

Figure 2.2 Testing Morgan's hypothesis that the recessive white-eyed mutation is on the female sex chromosome (sex-linkage). Male and female are indicated with the usual symbols, white eyes shown as white dots, red eyes as black dots. According to the hypothesis, the white-eyed mother has two X chromosomes, each of which carries the mutant allele, shown as a white dot. The red-eyed father has one X chromosome, which carries the wild type allele, shown as a black dot, and one Y chromosome which has neither allele. The hypothesis predicts that because female offspring must get one X chromosome from the father they will all be red-eyed. Conversely, because male offspring can only get their X chromosome from the mother, they will all be white-eyed.

Within a linkage group, however, alleles at different loci are likely to be inherited together. Each pair of alleles in a linkage group has a specific strength of linkage – a specific probability that they will be inherited together. Linkage maps arrange loci on the basis of how likely it is that alleles at those loci will be inherited together. Distances on these maps are measured in 'centimorgans' (Cm) – if two alleles are one centimorgan apart on a linkage map, then there is a 1 per cent probability that they will be separated in a single generation, or, put another way, they will be separated in 1 per cent of the individuals in the next generation.

Solid acceptance of the chromosome theory was the result of linking results in genetics to results in cytology, the observational analysis of parts and processes within cells. For example, some genetic phenomena could be related to observable errors in the replication of chromosomes during meiosis. Most significant, however, was the ability to correlate linkage maps with physical maps of chromosomes. In some cells, multiple copies of the same chromosome are produced and in some instances these remain attached all along their lengths to create a giant, 'polytenic' version of the chromosome. It is believed today that the function of this arrangement is mass production of certain gene products. In the 1930s, students of Morgan were able to make drawings of polytene chromosomes in *Drosophila*. These revealed a distinctive pattern of banding along each chromosome. The patterns made the exchange of segments between chromosomes directly observable. As well as standard crossing over (Figure 2.3), translocations, in which segments are exchanged between non-homologous chromosomes, and inversions, in which a segment of a chromosome is reinserted in the opposite orientation, can be documented using these banding patterns. The ability to relate these physical rearrangements to changes in the statistical linkage relationships between genes firmly established the idea that genes are distributed along chromosomes and that linkage is, very roughly, physical proximity of genes on a chromosome.

The clash between the leading German geneticist, Richard Goldschmidt, and his contemporaries in the 1940s and early 1950s provides further insight into the nature of classical Mendelian genetics. The success of the Morgan school in determining the linear order of genes on chromosomes allowed the discovery of 'position effects' in which changes in the relative position of genes on a chromosome are associated with changes in their phenotypic effects. This raised questions concerning the nature of mutation. Today we define a mutation as any change in the nucleotide sequence of the DNA in a chromosome. A mutation may occur either by the substitution, deletion, or insertion of one DNA nucleotide or by the translocation or inversion of a chromosome segment. In classical genetics, however, mutation was defined as a change in the intrinsic nature of an individual gene manifest in a heritable difference in phenotype. Mutations were thus distinguished from position effects, in which an intrinsically identical gene has a different effect because it has changed its location. Goldschmidt challenged this distinction. There was no direct evidence that chromosomes have distinctive structural parts corresponding to individual genes. Goldschmidt suggested that 'mutations'

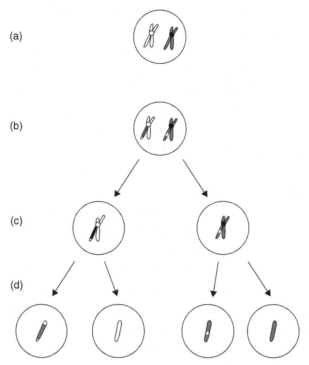

Figure 2.3 Schematic representation of meiosis for a single chromosome pair. (a) Diploid organisms like humans have two copies of each chromosome (homologous pairs of chromosomes). Before meiosis occurs the DNA in each chromosome is replicated so that each now consists of a pair of identical chromatids joined together at the centromere. (b) In the process of crossing over or homologous recombination one or more sections of a chromatid are exchanged for sections of a chromatid from the homologous chromosome. (c) The cell now divides into two daughter cells, each of which contains two copies of one of the homologous pair of chromosomes. This first cell division produces diploid cells with the standard number of chromosomes for this organism. It resembles the mitotic cell divisions by which organisms grow, except that crossing over is very rare in mitosis. (d) Meiosis concludes with a second cell division. This division is not preceded by DNA replication and the formation of chromatids. So it produces haploid cells with half the diploid number of chromosomes. These haploid cells give rise to the gametes (sperm and eggs or pollen and ovules).

and 'position effects' were simply smaller and larger changes in the physical structure of the chromosome. Because chromosomal changes on very different scales were known to have phenotypic effects, Goldschmidt argued that chromosomes probably contained a hierarchy of units of function. Famously, he denied that 'genes' exist, by which he meant that no unique structural unit corresponded to the unit of function in classical Mendelian genetics. Effectively, Goldschmidt was insisting that both aspects of the dual identity of the classical gene converge on a single unit – the material gene must correspond to the instrumental unit of genetic analysis. If this is not the case, Goldschmidt argued, then genes do not really exist. But Goldschmidt's views were completely unacceptable to most of his contemporaries (Dietrich 2000).

Some contemporary biologists argue that Goldschmidt has been vindicated by developments in molecular biology (Dillon 2003). The relationship between the physical structure of the chromosome and the physical gene as it is conceived today does indeed resemble Goldschmidt's picture of it, as we will describe in later chapters. But Goldschmidt was misguided to demand that geneticists abandon the idea of discrete genes occupying loci on chromosomes, because research into the physical basis of the gene was only one aspect of genetics. The Mendelian gene was not merely a hypothetical physical entity. As we have described, the Mendelian gene had a second identity as an instrumental entity. Distinguishing mutations from position effects was essential if mutations were to be grouped into distinct sets of alleles for the same locus. This in turn was essential to the practice of genetic analysis. Goldschmidt may have anticipated the future direction of molecular genetics, but it was continuing work in the classical, Mendelian tradition which made molecular genetics possible, as we will discuss below. Moreover, as later chapters make clear, the gene retains its identity as an instrumental unit today. Genetic analysis, and a version of the classical, Mendelian conception of the gene are alive and well.

We have argued that classical geneticists were not seeking to test the basic principles of Mendelism. Instead, they sought to solve specific puzzles about the genetic structure of traits in accordance with those principles. This description is somewhat reminiscent of Thomas Kuhn's (1962) description of 'normal science'. According to Kuhn, the history of science is punctuated by periods of massive intellectual upheaval or 'scientific revolutions'. The majority of scientific activity, which Kuhn called 'normal science', occurs in-between these revolutionary episodes. In periods of normal science the aim of research is not to question the basic framework established in the last scientific revolution, but to use that framework to answer detailed questions about the relevant scientific subject. It is assumed that the answers to these questions will reinforce and extend the basic framework that has already been accepted, not overthrow it. According to Kuhn, the successful researchers in periods of normal science are not those who challenge established theories, but those who show that apparent challenges to the existing framework can be resolved without questioning its fundamentals. Kuhn's description of the scientific process has been extremely controversial, and it is probably fair to say that while many scientists use the language of 'scientific revolutions' and 'paradigm shifts' few use them in accordance with Kuhn's original intent. Still fewer would accept the radical implication, drawn by some of Kuhn's followers, that scientific revolutions are not a rational response to the accumulation of evidence but instead resemble a religious conversion. Nevertheless, Kuhn's idea of 'normal science' is a useful corrective to other qualitative descriptions of science which make it seem that the confirmation or falsification of major theories is the daily business of science.

Classical genetics can certainly be interpreted as the use of theoretical and experimental ingenuity to show that even those patterns of inheritance apparently most inconsistent with Mendelian expectations can be explained by a suitable combination of Mendelian factors. The key to the ability to construct such explanations is the distinction between genotype and phenotype. By introducing that distinction, Johannsen freed genetics from the idea that the observable traits of organisms should behave in an obviously Mendelian fashion. How Mendelian heredity will appear at the level of the phenotype depends on the relationship between genotype and phenotype. By recognising more complex relationships between genotype and phenotype, geneticists were able to explain and predict very complex patterns of phenotypic inheritance without departing from Mendelian principles at the level of the genotype.

Genetic analysis of a character begins by postulating one or more loci which affect the character, and two or more alleles at each of those loci. One of the first characters to be subject to genetic analysis was the colour of flowers in the garden plant *Antirrhinum majus* (Snapdragon). The early Mendelian geneticist Muriel Wheldale began by dividing the colour of the flower into two distinct characters, the colour of the 'lips' of the flower and the colour of the 'tube' leading up to the lips. She postulated four loci, each with two alleles. One locus affected the colour of both tube and lips, two affected lip colour alone, and one affected tube colour alone. Genetic analysis also specifies the dominance relations between the alleles at each locus. Dominance may be of the simple kind in which one allele is dominant and the other recessive, or it may be more complex. For example, if crossing a homozygous red-flowered plant with a homozygous white-flowered plant produces a heterozygous pink-flowered plant, then the red allele is said to display 'incomplete dominance'. The alleles at different loci may exhibit some degree of genetic linkage, as we saw with Morgan's work on the white-eyed mutant. Genetic loci also interact by 'epistasis'. Two loci interact epistatically when the phenotypic effect of the genotype at one locus depends on the genotype at the other locus. Although the term epistasis had not been coined in 1907, Wheldale recorded a number of epistatic interactions between the four loci she postulated in *Antirrhinum*. For example, the presence of two recessive alleles at her first locus blocked any effect of alleles at her other three loci: these plants were white no matter what alleles they had at the other loci affecting colour.

By carefully defining characters, and postulating suitable combinations of loci and alleles with suitable relations of epistasis between the loci and suitable relations of dominance between the alleles, early geneticists like Wheldale were able to explain and predict complex patterns of phenotypic inheritance. Later geneticists added complexity to the genotype-phenotype relationship when they discovered position effects, as described above. Finally, the ideas of 'penetrance' and 'expressivity' helped to fit the phenotypic data to the underlying genotypic model. Penetrance measures the *proportion* of individuals with a given genotype who express a phenotypic marker of that genotype. It recognises that genotype and phenotype may be related probabilistically rather than absolutely. Expressivity measures the *degree* to which a phenotype appears: some people with a genotype associated with a disease may have the disease mildly, while others have it severely. Penetrance and expressivity may appear to be unscrupulous devices for fudging the data, but in fact they follow directly from the underlying rationale of the genotype/phenotype distinction. When he drew that distinction, Johannsen (1911) stressed that phenotypes themselves are not inherited. They result from the interaction of what is inherited – the genotype – and the environment of the developing organism. The phenotype gives the geneticist a signal about the genotype, but this signal is mediated by the process of development. Penetrance and expressivity reflect the noise in this signal introduced by variation in the other factors which affect development.

2.3 Beyond the Mendelian gene

The story of classical genetics departs from Kuhn's description of 'normal science' in an important way. The practice of classical genetics eventually led to the emergence of a new, molecular conception of the gene. But this did not occur because of dissatisfaction with the Mendelian paradigm and its rejection through a 'scientific revolution'. The gene had been a hypothetical physical unit of heredity since the earliest days of Mendelism. Increasingly sophisticated forms of genetic analysis provided grounds for a theory of the gene as a material unit of heredity. The gene in its identity as an instrumental unit of analysis helped to throw light on the gene in its other identity as a material unit. In the ensuing decades of genetics this second identity became the most prominent. But, as we will see in later chapters, the Mendelian 'paradigm' – the core ideas of Mendelian genetics and the practice of genetic analysis – were not overthrown, but persist even today.

In classical Mendelian genetics the gene played three theoretical roles. It was the unit of mutation – changes in genes give rise to new, mutant alleles of the same gene. It was also the unit of recombination. Changes in linkage relationships either separate genes which were previously linked, or link genes which previously segregated independently. Finally, the gene was the unit of function. The genotype which interacts with the environment to produce the phenotype is a collection of genes, and any effect of genotype on phenotype can be traced back to some gene or combination of genes. It was natural to project these ideas from the practice of Mendelian genetics onto the gene as a hypothetical material entity. It was expected that, when it was finally revealed, the physical gene would also be a unit of mutation, of recombination, and of genetic function. But the new, molecular concept of the gene that emerged in the 1950s did not entirely live up to these expectations. As we will explain in the following paragraphs, technical developments in genetic analysis allowed much more detailed maps of the chromosome ('fine structure mapping'). The results of this enhanced form of genetic analysis were interpreted in the light of the newly discovered biochemical structure of DNA. The new conception of the physical, molecular gene that eventually emerged from these advances was one in which the physical gene is only the unit of function, and not the unit of mutation or of recombination.

Recombination in classical Mendelian genetics means that a certain allele of one gene and a certain allele of another gene occur on the same copy of a chromosome, when before they had always been on different copies of that chromosome. This implied that genes themselves were the unit of recombination. This conception of recombination was reflected in a technique of genetic analysis known as the *cis-trans* or 'complementation' test. If two mutations are identified which affect the same character – for example, two mutations which both affect flower colour in a plant – the question arises whether they are two mutations in the same gene, or mutations in two separate genes. This can be assessed with the *cis-trans* test. The *cis-trans* test depends on the fact that in diploid organisms like fruit flies, snapdragon plants, or humans, chromosomes come in pairs. With the exception of the sex chromosomes X and Y, humans have two copies of each of their other twenty-two chromosomes. It is these pairs which exchange segments in crossing over during meiosis (Figure 2.3). The two chromosomes in a pair are said to be 'homologous' to one another. Two mutations are in *cis* position if they are both on the same member of a pair of homologous chromosomes: one piece of DNA contains two changes. Mutations are in *trans* position if one is on each of two homologous chromosomes: two pieces of DNA contain one change each.[1] If two mutations exist in *cis*, there is no way to tell whether they are mutations in the same gene or in two different genes. But if they exist in *trans* the two possibilities can be distinguished by the *cis-trans* test (Figure 2.4).

The test relies on the fact that most mutations are recessive. Only organisms in which both members of a pair of homologous chromosomes carry the mutation will show the mutant phenotype. To conduct the test two breeding lines of the organism are required, one of which is homozygous for first mutation, but free of the second mutation, and the other homozygous for the second mutation, but free of the first. When these two strains are crossed, offspring receive one of each pair of chromosomes from each parent. So they will receive only one copy of the first mutation, and only one copy of the second mutation. Now suppose the two mutations are in different genes. The organism has two copies of each of these genes, one mutant copy of each and one normal copy of each. Since both mutations are recessive, the organism will appear phenotypically normal. But suppose the two mutations are in a single gene. The organism will have two copies of this gene, one with the first mutation and the other with the second mutation. So both copies of

---

[1] The Latin terms *cis* (on the same side) and *trans* (on the opposite side) are used in subtly different ways in different areas of genetics and will be redefined in later chapters.

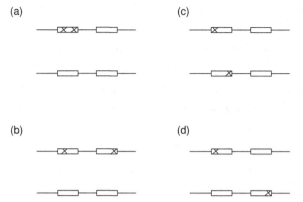

Figure 2.4 The *cis-trans* or 'complementation' test. Two mutations are known which affect the same phenotype. We want to know if these are two mutations in the same gene (a) or mutations in two separate genes (b). When the mutations are in *cis* position (on the same chromosome), this cannot be determined. However, by crossing an organism that carries only the first mutation with one that carries only the second mutation, we can create organisms that have the two mutations in *trans* position, on separate chromosomes (c and d). If the two mutations are in the same gene, both copies of that gene will be damaged and the organism will show the mutation (c). If the two mutations are in different genes, the organism will have one good copy of each gene and will not show the mutant phenotype (d). In this case the two mutations are said to 'complement' one another.

the gene will not function normally and the organism will be a phenotypic mutant. Thus, crossing two mutant lines to produce offspring with the two mutations in *trans* position and seeing if the two chromosomes 'complement' one another to restore normal function tests whether they are in the same gene.

But the *cis-trans* test does not work in this straightforward way if genetic recombination (Figure 2.3) can occur *within* a single gene. If parts of genes, as well as whole genes, can recombine, then recombination can stick together parts of two different copies of the same gene. If this happens with two different mutant alleles of the same gene, then some lucky organism may receive a copy of the gene which recombines the undamaged portion from one mutant allele with the undamaged portion from the other mutant allele, and is thus restored to normal function. Because this process relies on recombination occurring in exactly the right place, however, such intra-genic recombination will be very rare and very large-scale breeding experiments will be required to detect it with any degree of reliability.

Evidence of apparent intra-genic recombination can be explained in two different ways. One way is to multiply the number of genes that are believed to exist, so as to maintain the original significance of the *cis-trans* test. If two mutations, previously thought to be different alleles of the same gene, are able to recombine, we can insist that the two mutations must be in different genes, by definition, because the gene is the unit of recombination! This response was embodied in the idea of 'pseudo-alleles'. Pseudo-alleles were conceived by geneticists in various ways, but the general idea was that they are alleles of genes which are so close together on the chromosome that they appear to be alleles of a single gene unless subject to particularly careful analysis. But another response is to allow that there are multiple possible sites of recombination within a single gene, giving up the idea that the gene is the unit of recombination.

The resolution of this issue required a massive increase in the statistical power of genetic analysis. Successful genetic analysis requires a carefully developed experimental system – an organism with suitable general biological properties, such as a short generation time and good phenotypic markers of genotype, and strains of that organism whose genotypes have been constructed so as to allow clear tests of genetic hypotheses. Classical genetics first developed multi-cellular experimental systems, such as the fruit fly and maize, followed by fungal experimental systems such as the mould *Neurospora* and yeast *Saccharomyces* which were more suited to the analysis of biochemical phenotypes. Bacterial experimental systems were developed in part because their extreme simplicity allowed genetic analysis of the structure of individual genes. Attention also turned to the bacteriophages – viruses which live by attacking bacteria. The use of vast numbers of these viruses allowed reliable detection of very rare genetic events.

The geneticists of the 1930s had integrated their results with cytological observations to establish that genes are located on chromosomes. By the 1950s it had been established that it is the DNA molecule in each chromosome, rather than any of the protein molecules, that constitutes the genetic material of the chromosome. In 1953 James Watson and Francis Crick had published a plausible chemical structure for the DNA molecule, as we will explore in the next chapter. The task was now to relate the structure of the genetic material to the established facts of genetics. Working with bacteriophages from 1954 to 1961 Seymour Benzer was able to increase the resolution of genetic analysis and map the location of different mutations in such detail that each site was little larger than a single nucleotide of the corresponding DNA molecule. He used a version of the *cis-trans* test to map the units of genetic function – the regions of the genome which complement one another and restore normal function when they are present in *trans*. Bacteriophages do not have pairs of homologous chromosomes: their genome is a single strand of RNA or DNA. The bacteriophage analogue of the *cis-trans* test is whether two different mutant viruses, each of which lacks the genetic activity necessary to parasitise a bacterium, can parasitise it together. The two genomes 'complement' one another, or fail to do so, in a manner analogous to the two chromosomes in the original test. Because it was possible to use vast numbers of bacteriophage in a single experiment, it was possible to detect even extraordinarily rare events of intra-genic recombination. Benzer demonstrated that the functional unit defined by the *cis-trans* test (the 'cistron') consists of many sites at which different mutations can occur ('mutons'). Recombination could apparently occur between almost any two 'mutons', making the unit of recombination (the 'recon') the same size as the unit of mutation.

The terms 'muton' and 'recon' have not lasted. Mutation is something that happens to the DNA molecule itself, rather than to any genetically meaningful unit. Recombination is a highly regulated process involving chromosomes and an accompanying cast of enzymes. In analysing this process there is no need to divide the DNA itself into 'units of recombination'. But the term 'cistron' was widely adopted and is still sometimes used today. However, it is more common to simply talk of 'genes'. The basic molecular conception of the gene, the subject of Chapter 3, is the structural unit of DNA which performs the function which defines a cistron.

2.4 What was a gene?

The story told in this chapter can be told as one in which the molecular gene was glimpsed only dimly through the dark glass of breeding experiments, and progressively investigated until its true molecular nature was revealed. Many earlier philosophical analyses of genetics tried to understand it in this way. An older, less adequate theory called Mendelian genetics was reduced to a newer, more adequate theory called molecular genetics, just as Newton's theory of gravity was reduced to Einstein's theory of space-time. However, as we will explore in the next chapter, genetics poses many difficulties for traditional models of theory reduction. Moreover, as we will also see, the Mendelian gene as a concept implicit in the methods of genetic analysis is alive and well.

The idea that the old Mendelian gene was replaced by a new molecular gene would also fit nicely with some popular ideas in philosophy of science about the meaning of theoretical terms. Every philosopher is familiar with Hilary Putnam's account of how terms introduced in ignorance can nevertheless refer to the 'natural kinds' eventually revealed by science. Water turned out to be HOH because, without knowing this, it was HOH that people were drinking, bathing in, and calling 'water' (Putnam 1975). A parallel treatment of the word 'gene' would imply that when Mendel talked of 'factors' and Johannsen introduced the term 'gene' they were both, without knowing it, referring to the molecular gene.

The Mendelian and the molecular gene might also be fitted into the popular philosophical framework of 'role and occupant' (Lewis 1966). Some concepts can be understood in terms of the causes and effects of the thing being conceptualised, its 'causal role'. Lightning was originally known only as something which happens in thunderstorms, which causes bright flashes in the sky, and whose destructive effects we see as lightning strikes. For a concept like this, when it is discovered there is some concrete entity which occupies this causal role, then it follows that the concept picks out that concrete entity – lightning is the same thing as atmospheric electrostatic discharges. The gene was certainly originally identified by the causal role it played – it caused Mendelian patterns of inheritance. Later it was discovered that this causal role was occupied by pieces of DNA passing from parent to offspring. It follows necessarily that these pieces of DNA are Mendelian genes. But it does not follow that the Mendelian gene stands to the molecular gene as role to occupant.

The problem with both of these philosophical frameworks is that they lack any apparatus to recognise how the Mendelian gene is anchored in the experimental practice of genetic analysis. The molecular gene can only take over the role of the Mendelian gene if it can take over its role in genetic analysis. But, as we will see in the next two chapters, there are many pieces of DNA which play the role of Mendelian genes, but which are not molecular genes. In the case of the human genome, less than 2 per cent of the DNA sequence consists of coding sequences – the more complex, modern versions of cistrons – and thus of molecular genes. But much of the rest of the DNA sequence yields to genetic analysis: it has alleles that occupy loci, and therefore must contain Mendelian genes. So the molecular gene represents only one way to fill the role of the Mendelian gene. The molecular gene cannot be redefined to apply to any piece of DNA which can act as a Mendelian gene because it is anchored in experimental practices of its own, practices which rest on the linear correspondences between biomolecules which, as we will see in Chapter 3, are at the heart of what it is to be a molecular gene.

Rather than seeing the Mendelian gene as a primitive precursor to the molecular gene, we think a better account is one which recognises that the gene always had two identities, and that it retains those identities even today. There is no ontological mystery here – both identities are anchored in facts about DNA. But it does suggest a form of anti-reductionism, in that there is more than one scientifically useful way to think about DNA. This idea is developed at greater length in the next chapter.

Further reading

Robert C. Olby's *The Origins of Mendelism* (1985) remains the definitive account of the development of Mendelian genetics. For a broader perspective on the context in which Mendel's ideas took on such significance, see Müller-Wille and Rheinberger (2012). The idea of the dual identity of the gene is due to Raphael Falk (1986, 2000). The idea that classical genetics was primarily an experimental approach to understanding living systems, rather than a theory of heredity, can be found in Falk (2007, 2009) and also in a succinct presentation by C. Kenneth Waters (2004). For both the idea of *Drosophila* as an experimental system and a rich account of the development of that system, see Robert Kohler's *Lords of the Fly* (1994). Kenneth F. Schaffner gives a clear account of thirty years of attempts to reconcile the history of genetics with philosophical models of theory reduction (Schaffner 1993, Chapter 9).

3 The material gene

3.1 Introduction

In this chapter we describe the elucidation of the basic structure and function of DNA. This represented the successful conclusion of the search for the gene as a material unit of heredity. However, the causal role of the gene as it had been envisaged in classical genetics was very substantially revised in order to fit what had been discovered about the material basis of heredity. So the result of the molecular revolution in genetics was not that a previously defined causal role (the Mendelian gene) was filled by a newly discovered material occupant (the molecular gene). The molecular gene had a new role of its own, very different from that of the Mendelian gene. This explains some of the difficulties encountered by philosophers who have tried to explain how the Mendelian gene was reduced to molecular biology. Although the new, molecular identity of the gene was now its dominant identity, the old instrumental identity did not simply go away. The original role of the Mendelian gene continues to define the gene in certain areas of biological research: namely, those intellectually continuous with classical genetic analysis. We will show later in the chapter that it is sometimes necessary to think of genes as both Mendelian alleles and molecular genes, even when those two identities do not converge on the same pieces of DNA. The Mendelian gene turned out to be grounded in DNA, but the development of genetics has left us with more than one scientifically productive way of thinking about DNA and the genes it contains.

A major theme of this chapter is the emergence of 'informational specificity' as the key property of the molecular gene. The search for the molecular gene was the search for the source of biological specificity – the ability of biomolecules to catalyse very specific chemical reactions. We describe how specificity was transformed from a physical concept based on stereochemistry – the three-dimensional shape of molecules – to an informational concept based on the linear correspondence between molecules, most famously in the case of the genetic code.

The chapter closes with an exploration of the important philosophical question of whether Mendelian genetics was 'reduced' to molecular biology. We will argue that molecular biology is indeed a reductionist discipline: it shows that there is nothing more to genetics than the molecular processes it reveals. The traditional picture of an older theory being reduced to a new theory which can do all the old theory could and more does not fit the case of Mendelian and molecular genetics, however. So we examine whether molecular biology is 'reductionist' in various other senses. We conclude that the way in which molecular biology is an example of successful reductionist research is best captured by the 'neo-mechanist' account of reduction as the elucidation of underlying mechanisms associated with authors such as William Bechtel (2006, 2008). Following these neo-mechanist philosophers we argue that mechanistic explanations include both a reductionist phase which identifies and characterises the constituent parts, and an integrative phase that explains the specific ways in which those parts are organised so as to produce the phenomenon to be explained (Bechtel and Abrahamsen 2005; Craver and Bechtel 2007). Thus, molecular biology also bears out some traditional themes of anti-reductionism.

3.2 Molecular genetics

> Molecular biology was born when geneticists, no longer satisfied with a quasi-abstract view of the role of genes, focused on the problem of the nature of genes and their mechanism of action. (Morange 2000, 2)

The material nature of genes was not a question that could be answered via genetic analysis, and the pursuit of genetic analysis did not require it to be answered. Classical geneticists had no shortage of questions which could be addressed, very often successfully, with a purely instrumental notion of the gene. Some classical geneticists, however, notably Morgan's student Herman J. Muller, thought that genetics was fundamentally incomplete as long as the gene remained an unknown physical entity localised to a chromosomal position using indirect evidence. When the physical nature of the gene was uncovered, it would reveal how genes were able to replicate themselves (autosynthesis), to produce products which influence the phenotype (heterosynthesis), and to mutate (Muller 1947).

The search for the material gene was epistemically important because it showed a commitment to finding an epistemic pathway to the gene that bypassed the observed effect of the gene on the phenotype. When this commitment started to bear fruit it became possible to advance ideas about the gene which abandoned some of the commitments required if genes were to be epistemically accessible via genetic analysis. Features of the gene that previously could not be meaningfully called into question, and which were therefore effectively part of the definition of the gene, became features that could be tested and potentially rejected.

The material nature of the gene was progressively revealed by the new discipline of biochemistry which emerged in the first decades of the twentieth century. One aim of this discipline was to understand the agents of biological specificity – organic molecules that interact only with a very specific class of other molecules and thus enable the very precise chemistry required by living systems. From the mid-1930s it became increasingly clear that the specificity of organic molecules to their substrates (the molecule that is acted upon) is the result of stereochemistry: the complementary conformation of the molecules and their weak interactions through, for example, hydrogen bonds based on electrostatic attraction. The conformation of a molecule is its three-dimensional shape, which determines whether specific sites on molecules can come together. The interactions between those sites are much weaker than the covalent bonds of standard inorganic chemistry based on shared electrons, so that interactions between, and the conformation of, individual molecules can be altered by relatively low energies. While these principles turned out to apply to the structure and functioning of all forms of life, it was a special kind of protein – enzymes – that became the paradigmatic example of stereochemical specificity, or stereospecificity, through their highly specific catalytic action (Morange 2000, 15). The concept of specificity slowly came to be applied to the relationship between genes and their products, as well as to the relationship between enzymes and their substrates.

If the activity of the cell is explained by the stereochemical specificity of biomolecules it is natural to suppose that the effects of genes on phenotypes are mediated by the production of biomolecules with appropriate specificity. The little that was known about DNA suggested it was a monotonous molecule, consisting of repeating units of four nucleotide bases in equal proportions. It was believed to have little biological specificity, and perhaps to fill a structural role in the chromosome. However, in the 1940s it was shown that DNA alone could act as a 'transforming factor' which turned a non-virulent form of bacteria into one that caused disease. But the impact of this finding was delayed for some years. This was in part due to a reluctance to apply genetics to bacteria, for which no nucleus has yet been discovered and which seem to lack both chromosomes and sexual reproduction, in part due to the resistance to thinking that genes could be made of passive nucleic acids rather than active proteins.

Historians of science have stressed the very substantial changes in approach produced by the influx of scientists trained in physics into biology during the 1940s. These changes brought genetics and biochemistry closer together and paved the way for the molecular conception of the gene that prevailed from the 1950s to the 1970s. One of these former physicists, Max Delbrück, was convinced that understanding heredity would require a physical approach and an organism so simple that it could be conceived as a naked gene (Sloan and Fogel 2012). Bacteriophages – viruses which infect bacteria – presented themselves as organisms in which life was reduced to nothing more than self-replication, and was thus deemed perfect to study the physical basis of heredity 'without opening the biochemical "black box"' (Morange 2000, 45). The 'phage group' founded by Delbrück helped to establish the bacterial genetics which produced most of the key discoveries in the early decades of molecular biology. In 1952 this group demonstrated that bacteriophages infect the bacterium with DNA rather than with protein.

According to biochemist and historian Michel Morange the two require-ments for the acceptance of DNA as the genetic substance were, first, the separation of the question of the structural nature of genes from the problem of the mechanical characterisation of their function, and, second, decoupling the concept of specificity from its strong association with proteins (enzymes). Physicists were more likely to decouple the question of structure from func-tion than geneticists or other biologists. They were also more inclined to think about genetics in terms of information transfer and carriers of information (Morange 2000, 37).

Once DNA was perceived to have a central role in genetics researchers began to elucidate its structure. This is not the place to retell the story of the discovery of the DNA double helix by Francis Crick and James Watson in 1953. One important precursor, however, was biochemist Linus Pauling's attempt to model the structure of DNA. Pauling had shown the importance of the helix as a macromolecular structure, and also shown the importance of the structural constraints on the possible conformations of macromolecules imposed by the stereochemistry of peptide and hydrogen bonds between their constituent molecules. As a result, after first placing the nucleotide bases at the outside of the sugar-phosphate chain and then coupling like nucleotide bases with each other, Watson and Crick finally proposed the successful model of a double helix. It was characterised by the unique couplings of the bases adenine with thymine, and cytosine with guanine, that form the rungs of the ladder of which the sugar-phosphate chain provided the outer backbone (see Figure 3.1). This unique complementarity between the base pairs A-T and C-G suggested to Watson and Crick a possible mechanism behind some of the phenomena observed by classical genetics. In a famous understatement they wrote: 'This structure has novel features which are of considerable biological interest' (Watson and Crick 1953a, 737). The features of interest were a mechanism for self-replication, because the order of bases on each strand of the double helix specifies the order of bases on the other, and a mechanism for mutation, through the incorrect pairing of transient forms of bases (see Figure 3.1).

Many historians and philosophers of biology count this discovery as the moment when molecular biology gave rise to molecular genetics (e.g. Kitcher 1984). According to Sahotra Sarkar, the model of the double helix exemplifies the epistemic contribution of scientific models to theoretical unification, because it provided a point of contact between the different research programmes that were involved in the development of molecular genetics: classical genetics, biochemistry, biophysics, and a variety of theoretical considerations and deliberations. All had contributed constraints the gene must satisfy, and by satisfying all of them the 'confluent model' of the DNA double helix was 'standing at the confluence of four different research programs' (Sarkar 2005, 22). The classical gene must be both autocatalytic (DNA replication) and heterocatalytic (RNA transcription and translation), and able to mutate (base substitution or deletion). The biochemical gene must obey chemical principles (chemical properties of the DNA molecule) and be intimately related to enzymes (protein synthesis). The gene of 'biophysics', the name given at the time to some of the research out of which molecular biology developed, must satisfy certain bond length and stereochemical restriction between atoms (covalent bonds between sugar-phosphate groups and hydrogen bonds between nucleotides). Finally, in an influential work of theoretical biology Erwin Schrödinger (1944) had argued that genes must be

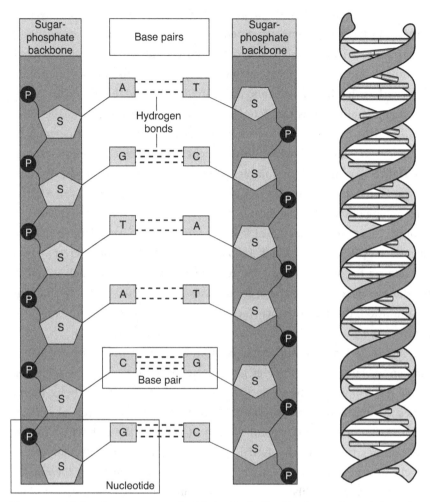

Figure 3.1 Molecular structure of DNA double helix. The figure shows base pairs formed from the nucleotides adenine (A) paired with thymine (T), and guanine (G) paired with cytosine (C), and the sugar-phosphate backbone. A fifth base, uracil (U), replaces thymine in RNA. (Reprinted from Ewa Paszek. 2007. Dogma of molecular biology. *Connexions*, 9 October. At http://cnx.org/content/m12382/1.5/)

simultaneously very stable (stable helical structure of DNA) yet infinitely variable (the sequence of DNA bases).

Many people have the impression that the discovery of the double helix coincided with the discovery of the genetic code. The conflation of these two discoveries is more or less standard in popular summaries of genetics. It is true that Watson and Crick's article on the implications of their proposed structure for genetics, published shortly after their first article, contained the hypothesis that 'the precise sequence of the bases is the *code* that carries the genetical *information*' (Watson and Crick 1953b, 965, italics added). But Morange argues that this merely reflected the widespread use of these two terms at that time, not a real appreciation of the implications of a genetic code. For the discovery of the structure of DNA to lead on to the elucidation of the main processes involved in the protein synthesis – transcription and translation – one important obstacle remained to be removed: the association of specificity with the structural conformation of proteins that was so entrenched in biochemistry (Morange 2000, 120f., 139).

Three years after Watson and Crick's model was published the molecular biologist Joshua Lederberg first used a term akin to 'informational specificity' but without really abstracting away from the deeply rooted, biochemical view that it is the physical form of a molecule, including that of DNA or RNA, that provides its specificity:

> The hypothesis which obviously underlies the one-to-one theory is that a gene works as a unique template for 'stamping the specificity' on an enzyme. My philosophical reservation is against the implication that 'specificity' (or 'information,' as it is called nowadays) is something apart from structure. (Lederberg 1956a, 167)

In the same paper Lederberg showed that his understanding of gene action was still wedded to the classical genetic idea that it determines the phenotype from a distance, and to the biochemical idea of specificity:

> The alternative, which I prefer, is that all the specifications are already inherent in the genetic constitution of the cell: the inducer signals a regulatory system to accelerate the synthesis of the corresponding enzyme protein.
>
> [...]
>
> With respect to the fundamental question of how the genes work, there is no debate at all that genetic functions are ultimately mediated by enzymes; we are discussing only the organizational details. (Lederberg 1956a, 161, 167)

However, in a different paper published in the same year, he wrote:

> The specificity of DNA is believed to depend on the sequence of these four alternatives; it is not surprising that very large molecules are required to store biological information of ultimate complexity in a language with such a simple alphabet. (Lederberg 1956b, 268)

There are several ways to interpret these statements: (1) Lederberg considers the relationship between nucleic acids and proteins to be stereospecific rather than structurally arbitrary and sequence-specific; (2) Lederberg may have been foreseeing the actions of transcription factors that bind to the DNA in order to recruit the transcriptional machinery; or (3) he may have reasoned that whatever genes do, any chemical reaction they are involved in would need the catalytic efficiency of enzymes (which turned out to be the enzyme polymerase and the ribosome; see below). This third possibility is exceedingly unlikely because it doesn't appear that he or other biochemists at that time distinguished between the *efficiency* of an enzyme or chemical reaction, and its substrate or sequence *specificity* (a distinction that will be scrutinised in more detail in 4.5).

3.2.1 The sequence hypothesis and the Central Dogma

It was Francis Crick who in his famous 'Central Dogma of molecular biology' and 'sequence hypothesis' made the transition away from stereochemical specificity to informational or sequence specificity:

The sequence hypothesis

In its simplest form it assumes that the specificity of a piece of nucleic acid is expressed solely by the sequence of its bases, and that this sequence is a (simple) code for the amino acid sequence of a particular protein.

The Central Dogma

This states that once 'information' has passed into protein *it cannot get out again*. In more detail, the transfer of information from nucleic acid to protein may be possible, but transfer from protein to protein, or from protein to nucleic acid is impossible. Information means here the *precise* determination of sequence, either of bases in the nucleic acid or of amino-acid residues in the protein. (Crick 1958, 152–3, italics in original)

In other words, the linear sequence of nucleotides in a segment of a DNA molecule specifies the linear sequence of nucleotides in an RNA molecule, and that molecule in turn determines the linear sequence of amino acids in a protein through 'informational specificity' – that is, via the genetic code whose details were to be elucidated in the early 1960s. In the remainder of the book we will refer to the sense of information which Crick defined above as 'Crick information', and it will prove to be very important.

The colinearity between genes and their products was at the heart of the new conception of the gene associated with the rise of molecular biology, the so-called 'classical molecular gene'. This is the way of thinking about genes still found in today's textbook definition: a gene consists of a 'promoter region' which acts as a signal to the machinery that transcribes the DNA into RNA, followed by an 'open reading frame', a series of codons each corresponding to an amino acid plus a 'start codon' and a 'stop codon' which act as signals to the machinery that translates the RNA into protein. The gene of molecular biology is the image of a gene product in the DNA[1] – a sequence of DNA bases which has a part-by-part correspondence to the product derived from it. The central epistemological role of linear correspondence between molecules in molecular biology was brought to philosophical attention by C. Kenneth Waters (1994). Linear correspondence between molecules is fundamental to biologists' ability to identify and manipulate those molecules. Linear correspondence is thus at the heart of the molecular conception of the gene: a linear correspondence to some molecule of interest picks out a certain sequence of DNA nucleotides as the 'gene for' that molecule:

> At its heart, a synthetic DNA probe is a rational, linear, digital signature to locate any counterpart in the analysand. Its core of combinatorial specificity can be contrasted to that of antibodies, which is founded on three-dimensional shapes of the immunoglobulin and its targets. (Lederberg 1996, 22)

Crick intended the Central Dogma and the sequence hypothesis as a theoretical framework to guide future research on protein synthesis and gene expression, rather than as truths beyond question, even though that was how the Central Dogma in particular has been interpreted and criticised. At the time, neither was based on solid experimental data, but rather on informed conjecture. Firm answers awaited the discovery of the intermediary molecules and subroutines involved in the processes of transcription and translation, such as messenger and transfer RNAs, the genetic code between nucleic acids and amino acids, the enzymes involved such as polymerase and ribosome, transcription factors, and regulatory DNA sequences. But the Central Dogma and sequence hypothesis certainly marked the beginning of a paradigm shift in genetics, and Crick's insights in his 1958 paper 'On Protein Synthesis' were

---

[1] We owe this very nice expression to Rob D. Knight.

nothing less than visionary. He introduced a new way of thinking about biological specificity, a way of thinking that underpinned not only a new conception of the gene, but also the new technologies that would flow from molecular biology.

3.2.2 The genetic code

The race to determine the genetic code that started in the mid-1950s was a race between theoretical physicists and mathematicians, on the one hand, and experimental chemists, on the other (Sarkar 2005). At issue was the correspondence between nucleic acid bases and amino acids. Crick proposed that the more than one hundred known amino acids in nature can be reduced to a basic set of twenty amino acids, and his list turned out to be correct. The search soon converged on triplets of three bases as the basic 'words' of the code, later called a codon. The three-base code had the advantage of providing enough possibilities, but posed the problem of how sixty-four combinations of the four nucleic acid bases correspond to only twenty amino acids. It turned out that the code is 'degenerate' in the sense that up to three different triplets of nucleotides can code for one and the same amino acid. Numerous theoretically driven solutions to the coding problem led to dead ends, and the prize fell to the biochemists who through painstaking in-vitro experiments discovered the first word of the code in 1961, and by the mid-1960s the complete code (Figure 3.2 and Table 3.1). Although it is degenerate, the code is never ambiguous: each codon specifies only one amino acid. The code turned out to be non-overlapping, in the sense that each codon ends where the next begins. The code also turned out to be comma-less, in the sense that nothing in the actual DNA sequence marks where one codon ends and the other begins: how the sequence is divided up into codons depends on where you begin (the 'reading frame'). The elucidation of the code proved once and for all that the conformation of messenger RNA molecules does not play any role in the specification of amino acids and hence in the synthesis of proteins: the sequence hypothesis was correct.

3.3 The classical molecular gene

With the unravelling of the genetic code and of the basic processes of transcription and translation in the 1960s the two identities of the classical gene,

		Second base				
		U	C	A	G	Third base
First base	U	UUU phe / UUC	UCU / UCC / UCA / UCG ser	UAU tyr / UAC	UGU cys / UGC	U / C / A / G
		UUA leu / UUG		UAA stop / UAG stop	UGA stop / UGG trp	
	C	CUU / CUC leu / CUA / CUG	CCU / CCC pro / CCA / CCG	CAU his / CAC	CGU / CGC arg / CGA / CGG	U / C / A / G
				CAA gln / CAG		
	A	AUU / AUC ile / AUA	ACU / ACC thr / ACA / ACG	AAU asn / AAC	AGU ser / AGC	U / C / A / G
		AUG met or start		AAA lys / AAG	AGA arg / AGG	
	G	GUU / GUC val / GUA / GUG	GCU / GCC ala / GCA / GCG	GAU asp / GAC	GGU / GGC gly / GGA / GGG	U / C / A / G
				GAA glu / GAG		

Figure 3.2 The genetic code: the sixty-four possible three-nucleotide codons and the twenty amino acids for which they code (three-letter abbreviated names).

Table 3.1 *List of the standard twenty amino acids. These form the structural units that join together to form the polypeptide chains which make up proteins*

Amino acid name – Three-letter code – One-letter code		
alanine – ala – A	glycine – gly – G	proline – pro – P
arginine – arg – R	histidine – his – H	serine – ser – S
asparagine – asn – N	isoleucine – ile – I	threonine – thr – T
aspartic acid – asp – D	leucine – leu – L	tryptophan – trp – W
cysteine – cys – C	lysine – lys – K	tyrosine – tyr – Y
glutamine – gln – Q	methionine – met – M	valine – val – V
glutamic acid – glu – E	phenylalanine – phe – F	

the instrumental Mendelian and the hypothetical material, seemed to have converged neatly on a single, well-defined entity – the classical molecular gene. The functional definition of the gene that underlay genetic analysis and the structural definition of the material gene had turned out to be two ways to pick out the very same things. Looked at more closely, however, the functional definition had been significantly revised so as to take account of findings about the material gene.

We saw in 2.3 how Benzer's work demonstrated that the gene was not the unit of recombination or of mutation. When Muller described the roles of the still largely hypothetical physical gene in 1947 he identified three such roles: self-replication, mutation, and the production of biochemical products which influence the phenotype (Muller 1947). The classical molecular gene, however, is not the unit of replication, which is the whole DNA molecule of which it is a part. Nor is it the unit of mutation, which is a single DNA nucleotide. The only role with respect to which the molecular gene is the unit of function is that of producing a product. So the functional role of the gene was reduced to this alone in order to fit the molecular structures that had been uncovered.

Furthermore, the concept of the gene in molecular biology was restricted to sequences that fulfilled this new role: not all segments of chromosomes that behave as Mendelian alleles count as genes under the new molecular conception. As we will see in more detail below, untranscribed regulatory regions not immediately adjacent to the coding sequences they regulate can segregate independently of those coding sequences, and so can function as separate Mendelian alleles, but they are not separate molecular genes.

The classical molecular gene was a highly successful example of the research strategy of identifying a functional role, searching for the mechanism that fills that role at a lower level of analysis, and then using knowledge of that mechanism to refine understanding of function at the original level of analysis. In this case, the original role was that of the Mendelian gene, which explained the phenomena of heredity by replicating, recombining, influencing the phenotype, and occasionally mutating. The molecular gene was intended to be the occupant of the Mendelian role, when in reality it only played one part of that role. But contrary to the usual philosophical picture of role-occupant analysis (see 2.4), the original role of the Mendelian gene continued to be important alongside the new role, and the new molecular occupant of that role.

3.3.1 The mechanism of protein synthesis

From around 1950 it was known that protein synthesis takes place in a cellular structure originally termed the microsome but later renamed the *ribosome*, meaning a cellular body formed out of ribonucleic acid particles. Before the discovery of messenger RNA many believed that ribosomes were the missing link between DNA and protein and that it was the ribosome which specified the final amino acid sequence. A very small RNA particle had also been found, first called soluble RNA and later renamed transfer RNA (tRNA). Transfer RNA vindicated an early speculation of Crick about an adaptor molecule mediating between amino acids and their nucleic acid template.

The French geneticists François Jacob and Jacques Monod revealed the role of messenger RNA (mRNA) in 1960 in collaboration with other researchers in the UK, USA and France. Experiments had shown that, while the sequence of DNA differs considerably between species, this is not true with respect to ribosomal RNA. This was a problem for the idea that the ribosome is the intermediary molecule between DNA and protein. Indeed, these findings 'began to shake the fragile edifice of molecular biology and to cause even ardent supporters to question the idea of a genetic code' (Morange 2000, 145). But Jacob and Monod, among others, had already speculated about the necessity of an intermediary between genes and ribosomal RNA. The majority of experiments in the young field of molecular genetics were conducted with bacteria (prokaryotes), unicellular organisms without a nucleus and without internal membranes that could spatially separate genes from ribosomes and proteins. Nevertheless, it was known from earlier cytological studies that in eukaryotes, organisms including fungi, plants, and animals, whose cells contain a nucleus and other complex, membrane-bound structures, the genetic material is found in the nucleus but ribosomes are located in the cytoplasm, which is where protein synthesis takes place.

This might not have been regarded as a problem by classical genetics, which was not concerned with the mechanism by which genes affect the phenotype, but the new molecular genetics was based on the idea that genes provide detailed specification for proteins. How could this occur when genes were separated from the site of protein synthesis by a membrane? The puzzle was resolved by the discovery that the ribosome only functions as structural and catalytic support for the production of proteins. The much smaller and shorter-lived mRNA is transported across the membrane and is translated into the amino acid chain. The idea that DNA is related to RNA production had been around for some time, but only now could it be shown that RNA was synthesised in a process called transcription that is analogous to DNA replication. A complementary strand of RNA is formed by one of the two DNA strands upon its separation by the enzyme RNA polymerase (instead of the

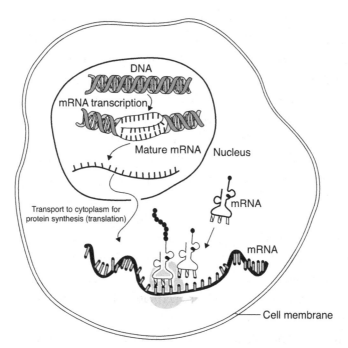

Figure 3.3 Protein synthesis in eukaryotes: simplified schematic description of the transcription of DNA to mRNA in the nucleus, and the translation of mRNA into a polypeptide chain in the cytoplasm. The RNA polymerase is not shown, but the ribosome (light grey) is shown translating the mRNA to a polypeptide with the help of tRNAs.

related enzyme of DNA polymerase which catalyses replication). The two main differences between the processes of replication and transcription are that each newly matched RNA nucleotide will detach itself from the DNA to form a single-stranded mRNA, and that the adenine is matched not with thymine, but with the closely chemically related uracil. The two main processes that allow the transfer of genetic information proposed by the Central Dogma were now in place: transcription and translation (Figure 3.3).

While one may often read that DNA encodes RNA, that is not strictly correct, since the genetic code is a relationship between RNA and amino acids. The relationship between DNA sequence and RNA sequence is dictated by chemical complementarity, rather than an evolved coding scheme.

3.3.2 The operon model

Classical Mendelian genetics did not try to explain embryonic development and differentiation. The existence of genes was inferred from the inheritance patterns of the phenotypic characters they were believed to determine. The developmental mechanisms that connected the genes to these phenotypic characters were unknown. Embryologists interested in this topic continued to support a major role for the cytoplasm in development, differentiation, and morphogenesis (Sapp 1987). Even Morgan, who had started out as an embryologist before he became one of the founding fathers of classical genetics, emphasised the interplay between the cytoplasm and the genetic material in the nucleus during differentiation. Some even hypothesised the independent continuity of cytoplasmic components which they termed 'plasmagenes'. But the rise of molecular genetics propelled the question of developmental mechanisms to the forefront of scientific enquiry. While morphogenesis may not be such an obvious problem when working with single-cell organisms, bacteria develop in the sense that they react to different environmental and life-cycle conditions and so they too require gene regulation.

In their famous 'operon' model Jacob and Monod (1961) proposed a general regulatory mechanism for all genes that function via a negative feedback mechanism able to repress gene expression after protein synthesis had commenced. An operon is a sequence of DNA containing a cluster of genes whose gene products are functionally related and which are under the common control of regulatory sequences called the promoter, a regulator gene with its own promoter, and operator sequence. The regulator gene synthesises a repressor which binds the operator, while the structural genes are under the control of the operator.

The operon model was based on the first known example, the *lac* operon in the bacterium *Escherichia Coli*. The *lacZ* gene (the first gene of the *lac* multi-gene region that also contains *lacY* and *lacA*) codes for the enzyme ß-galactosidase which breaks lactose into simpler sugars. Its transcription conveniently depends on the presence of lactose. A signal produced by the simultaneous presence of glucose, which is a more efficient energy source, can override the lactose stimulus. While the original model only assumed the existence of a negative feedback mechanism, the *lac* operon later turned out to also contain a positive control element. Two DNA-binding proteins, the *lac* repressor and the CAP activator, mediate between the two environmental signals lactose and glucose and the *lac* operon. Because RNA polymerase can spontaneously bind the promoter of *lac* there exists a low basal level of transcription even if neither lactose nor glucose, and hence neither binding protein, is present. If not overridden by the presence of glucose, lactose causes the CAP activator to bind the cap site within the promoter that activates

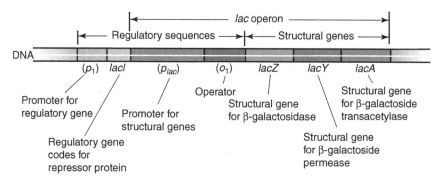

Figure 3.4 A model containing the positive and negative regulatory and coding elements in the *lac* operon.

transcription more efficiently than the usual background level. In the presence of glucose the repressor binds to the operator immediately upstream of the transcription start site, which prevents the polymerase from binding to the promoter with or without CAP being present, and transcription is stalled (Figure 3.4) (Ptashne and Gann 2002, 13ff.).

Regulatory regions are usually, but not always, located 'upstream' from the DNA region that contains the structural genes which they regulate. Locations on a DNA strand are divided between 'upstream' and 'downstream' with respect to the first base of the start codon that marks the transcription start site where mRNA will be transcribed. The position of that base is designated as +1 and is the first 'downstream' location. Upstream sequence locations are designated with numbers that start with a minus sign. DNA is always read and transcribed in the same direction, starting from the 5′ end and proceeding toward the 3′ end of the DNA strand.[2] The two strands of DNA are 'antiparallel', which means they are going in opposite directions. When an activator binds to the promoter sequence of the operon it recruits a transcriptional enzyme, the polymerase, that will proceed to transcribe a 'polycistronic' mRNA that contains the coding regions of several genes (or 'cistrons'; see 2.3). These will then either be translated as a single unit into several proteins or cut into separate functional domains before translation. In bacteria transcription and translation is often an ongoing process, with the two phases happening in close physical and temporal proximity to each other.

---

[2] These two ends are chemically different – the terms 5′ and 3′ derive from the biochemical labels for specific carbon atoms in the DNA molecule.

3.3.3 Regulated recruitment and combinatorial control

In the years after the initial research into bacterial gene regulation it turned out that even in these relatively simple organisms, gene regulation can be much more complex than the *lac* operon model. We will not elaborate here, but we will introduce a principle which applies to the *lac* operon model and many other gene expression mechanisms in prokaryotes, and which applies to literally all gene expression in eukaryotes. This is the principle of *regulated recruitment and combinatorial control*, which will become a familiar theme over the next two chapters. For example, an RNA polymerase that is constitutively active nevertheless needs to be actively recruited to the promoter site of a specific gene or genes by specific DNA binding proteins (generally called transcription factors) to ensure high-level and efficient transcription. Often in such cases of gene regulation, activator and repressor binding proteins work antagonistically, as in the case of the *lac* genes (Ptashne and Gann 2002, 13ff., 49). The fact that several different factors interact, sometimes reinforcing one another and sometimes antagonistically, allows the regulation of which genes are expressed, and also control of which products are expressed from those genetic loci. In eukaryotes these processes involve many more transcription factors and other associated proteins and regulatory RNA (which we will from now on call *trans*-acting factors)[3], and much longer and more complex regulatory regions upstream, downstream, and within the structural genes (which will be called *cis*-regulatory modules, regions, or factors). The expression of each eukaryote gene relies on the regulated recruitment and combinatorial control of a large number of *trans*-acting factors, and signals received from the environment that can control these *trans*-acting factors. The existence of regulated recruitment and combinatorial control,

> suggests that to evolve increasingly complex biological systems, it may not be necessary to invent many new kinds of gene products. Rather, more sophisticated functions can be achieved by, for example, increasing the number of interactions that any one protein can make, through the reiterated use of simple binding domains, thereby expanding the possibilities for combinatorial associations. (Pawson, in Ptashne and Gann 2002, xvi)

---

[3] The use of *cis* and *trans* here is similar to the last chapter. *Cis*-acting factors are on the same strand of DNA as the gene they regulate. *Trans*-acting factors are transcribed from elsewhere and transported to interact with the DNA strand they regulate.

The regulated recruitment of transcription complexes, and their combinatorial control of transcription, leads to a one-to-many relationship between a molecular agent and its effects in different contexts. One and the same transcription factor may act as an activator or repressor depending on which regulatory sequence it binds or which other factors it interacts with. There is a profound difference between the basic molecular function of a molecule, such as the stereochemical affinity of a transcription factor for a binding site, and the realised cellular function of that molecule, which it owes to its interaction to other entities. The informational specificity inherent in the coding sequence of a gene underpins specificity in its original sense of the ability of organisms to exercise exquisitely sensitive control over biochemical processes, but it does not do so in a simple, direct, and unambiguous way. How the inherent specificity of genes is put to work in particular contexts depends on regulated recruitment and combinatorial control.

3.3.4 Implications of Jacob and Monod's work on gene regulation

The implications of the work of Jacob and Monod went far beyond its experimental success in finally proving the existence and function of a messenger RNA. First, by finally elucidating the relationship between DNA and RNA they confirmed Crick's Central Dogma and therefore provided the 'final break between form and information', a necessary step for the maturation of molecular biology as a field in its own right (Morange 2000, 149).

Second, they demonstrated the existence of regulatory genes – those which code for transcription factor proteins – whose sole function it is to regulate the activity of other genes that code for enzymes or structural proteins. They revealed the first inklings of a hierarchical network of genes, foreshadowing the gene regulatory networks (GRNs) of which much more will be said in the following chapters.

Third, following directly from the preceding point, their model of the regulation of gene expression stressed the important regulatory function of many gene products, either by binding to the DNA itself to control gene expression or by interacting with a DNA-binding protein. Their existence added a feedback dimension to the simple linear figure drawn by Crick (DNA → RNA → protein), with an arrow for 'regulation' rather than 'transfer of sequence information' from proteins back to DNA. In the next chapter we will see that some of this regulation has some important consequences for the final sequence of the gene product, and hence we will argue that while you cannot transfer the same Crick information back from protein to DNA, some proteins do provide Crick information for other gene products. We will discuss the discovery of sequence-specifying factors other than the original DNA sequence in Chapter 4.

Fourth, Jacob and Monod's work drew attention to the function of some DNA sequences that are not repositories of Crick information for RNA or protein. Their function is not to provide sequence information as described in the sequence hypothesis and the Central Dogma, but to act as mediators in the activation of sequence information. We will see in Chapter 4 that there are many forms of functional non-coding DNA.

A fifth result of Jacob and Monod's work was to open the way to study the role of genes in development and differentiation. All cells in the body contain the same genes, and developmental biology is the study of how cells differentiate to produce the many different tissues that make up the body. Many years after the split between embryology and genetics, a split that despite all its negative consequences allowed for the birth and maturation of genetics, their work provided the first hint of a possible reconnection and even reconciliation between these two fields.

Finally, their model described a mechanism by which the organism is open to environmental influences, a feedback mechanism by which environmental signals could influence gene regulation. A fundamental truth that cannot be stressed often enough is that all organisms must respond to their environment.

3.3.5 The consolidation of molecular biology

By the mid-1960s many scientists thought that the major problems of molecular genetics had been solved, and were inclined to leave other investigators to iron out the details (Waters 2004; Falk 2007). But Monod's famous quip that whatever is true for the bacteria E. coli will be true for the elephant turned out to be premature, and forty years later it seems unlikely that molecular biologists will find themselves out of work anytime soon. The next decade was characterised less by groundbreaking discoveries than by an extraordinary expansion of molecular genetics and biology both by the founding of more laboratories and institutes and by swift takeovers of existing fields on which the molecular biologists imposed their vision of a new biology. That vision was described in terms of information transfer, code, message, and memory, metaphors that everybody understood (Morange 2000, 173f.).

In the light of all this, the discovery of the enzyme reverse transcriptase came as a bombshell. This enzyme is produced and used by retroviruses to convert their genetic material, RNA, into DNA in order to integrate it into the genome of the infected cell of their host. This allows them to hijack the host's protein synthesis machinery. This apparent reversal of the Central Dogma was perceived as a small revolution. Crick offered a quick clarification of his Central Dogma in 1970 that explicitly allowed information transfer from RNA to DNA. In the end this discovery, rather than shaking the foundation of the discipline, led to the new era of genetic engineering, since it provided a mechanism for inserting new genes into an existing genome. Most if not all tools of this new technology exploited what can be regarded as the most important breakthrough of molecular biology: namely, the discovery of the colinearity between the genetic material, its intermediary molecules, and the primary structure of its final products. This colinearity provides the means to isolate, characterise, manipulate, and modify genes.

3.4 The rediscovery of complexity: gene regulation in eukaryotes

The realisation that gene regulation is massively more complex than had initially been thought resulted from a shift in research from single-celled organisms without a nucleus, like bacteria, to eukaryotes (uni- and multi-cellular organisms with nucleus, cytoplasts, and additional structural features, such as diverse internal membrane systems). The main differences in gene regulation stem from this complicated architecture of the cell and a larger and more flexible genome catering to the demands of a more highly differentiated organism.

The first major difference lies in the structure of the cell. Eukaryotic cells are much larger and are divided into different compartments, the most important of which is the nucleus containing the chromosomes and which is isolated from the cytoplasm by an internal cell membrane. The cytoplasm itself is highly structured through a mass of other internal membranes and cellular organelles, some isolated through their own membranes, such as mitochondria and chloroplasts, and some particulate complexes, such as the ribosomes. These membrane-bound organelles, which are the energy producers of the cell, turned out to contain their own DNA and their own transcription and transla-tion machinery. Transcription and translation of the nuclear genes take place in two different compartments of the cell, which requires the introduction of an intermediary molecule that transfers the genetic sequence information between DNA and proteins from the nucleus. DNA gets transcribed into a pre-mature mRNA, at this stage a complete copy of the DNA. This pre-mRNA is a notoriously short-lived molecule because it immediately undergoes a whole range of post-transcriptional modifications (see below), partly to prevent the premature decay of the mRNA and partly to allow for its transportation out of the nucleus.

In multi-cellular organisms cells differentiate from a single zygote into cells with vastly different phenotypes as a result of the synthesis of different proteins. The mosaic theory of development proposed by August Weismann and Wilhelm Roux at the end of the nineteenth century, according to which different cell types receive different parts of the hereditary material, had been rejected by the beginning of the twentieth century. Studies in the 1960s confirmed that it was indeed not the DNA content that differed between differentiated cell, but the RNA content and in consequence the proteins that were expressed. From this it can be deduced that the differential regulation of gene expression is responsible for the production of the various cell types, and not the original sequence information within the genes. This insight focused attention away from sequence information to the processes responsible for its differential expression.

3.4.1 Transcriptional control

The genetic material of eukaryotes is longer and more densely packed than bacterial or viral DNA. DNA is wrapped tightly around a core of several histone proteins, the nucleosome, with which it forms the material chromatin which is further condensed to form the chromosome (see Figure 5.1 and our cover image). Chromatin renders DNA inaccessible to the transcriptional machin-ery; therefore eukaryotes have no default or constitutive transcription: all gene expression needs to be regulated. As a first step transcription factors must recruit a chromatin remodelling complex to cleave the DNA away from the nucleosome (these so-called epigenetic mechanisms are described in detail in 5.3). While transcriptional regulation in bacteria is a relatively simple and efficient affair, things are quite different in eukaryotes. For instance, with some notable exceptions eukaryotes do not use co-regulation of related genes via polycistronic transcripts as occurs in the *lac* operon. In eukaryotes the single most important regulatory mechanism at this stage of protein synthesis is the regulated recruitment of a large number of transcriptional factors and co-factors and their combinatorial control of RNA polymerase and associated factors (Ptashne and Gann 2002). A eukaryotic cell has three different kinds of RNA polymerase, each specific for different genes (either protein or RNA genes). In line with the many kinds of transcriptional factors involved in transcriptional activation, the regulatory sequences to which they bind are more diverse, complex, and widespread. There may be more than one promoter and associated regulatory modules upstream of the gene, and regulatory sequences can also be found within and even downstream of the gene. Very distantly located sequences, called enhancers and silencers, can bind to activators or repressors and help to recruit or to block the transcriptional machinery. Gene boundaries are defined by transcriptional start and termination sites, and there may be alternative sites available that are associated with different promoters.

3.4.2 Post-transcriptional regulation and translation

The product of transcription is a short-lived initial transcript called the pre-mRNA (also called primary or heterogeneous nuclear mRNA), which post-transcriptional processing will transform into a typically shorter mature mRNA. A cap is added to its 5′ end, and the 3′ end is extended by a poly-A tail that consists of up to 200 adenines, a process called polyadenylation. Both mechanisms enhance the stability of the mRNA and its transport into the cytoplasm. In the cytoplasm a ribosome will attach itself to the transcript and move along the mRNA, translating each codon of three nucleotides into their respective amino acids, with the help of transfer RNAs. Apart from four codons that function as start or stop signals for transcription, there exist sixty different tRNAs, each of which combines one particular anticodon with its specific amino acid. After each amino acid bonds with its neighbour it is cleaved from the tRNA, which subsequently leaves the ribosome (see again Figure 3.3).

3.4.3 Split-genes and introns

Since the 1970s further investigation has tended to undermine the idea that the functional role of the molecular gene – specifying the linear sequence of

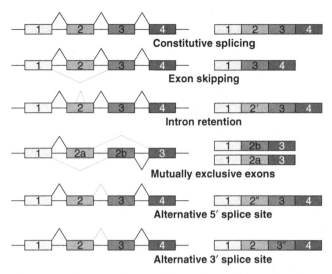

Figure 3.5 The main forms of alternative splicing. Constitutive splicing, where all exons are retained, and five forms of alternative splicing. Exon skipping skips exon 2. Intron retention treats an internal part of exon 2 as an intron. In mutually exclusive splicing, either exon 2a or exon 2b is retained but never both together. In alternative 5′ or 3′ splice sites an exon (here exon 2 and exon 3) can have two alternative splice sites at their 5′ or 3′ end, effectively cutting off part of the exon at the beginning or end.

elements in a gene product – is filled by natural units of molecular structure at the level of the DNA. The first surprise came when it was discovered that eukaryotic genes come in pieces: they are 'split-genes'. In eukaryotes the pre-mRNA is processed by cutting out large non-coding sequences, called introns, after which the remaining coding sequences, called exons, are spliced together to make the final mRNA transcript. This process is called splicing (or, more technically, *cis*-splicing, since a more complicated process of *trans*-splicing was later detected; see Chapter 4). The intervening sequences (introns) are often much larger than the remaining coding sequences (exons) that form the mature mRNA. Exons sometimes comprise just 5–10 per cent of the original DNA sequence. This kind of post-transcriptional regulation disturbs the perfect colinearity between gene and protein, which was the hallmark of the classical molecular gene concept. Having discovered splicing, biologists soon detected that alternative versions of mature mRNA transcripts can result from the cutting and joining of different combinations of exons, a process called alternative splicing (see Figure 3.5). While all eukaryotic pre-mRNAs need to be spliced, alternative splicing was for decades regarded as a rare complication. Now we know that almost all human genes are alternatively spliced, some into a large number of alternative splice forms. Alternatively spliced transcripts are then translated into different but mostly structurally related proteins (isoforms). Alternative splicing creates an obvious functional role for introns. They allow for a huge increase in the number of gene products from a limited number of genes, and they contain important regulatory sequences that bind splicing factors. Today the molecular gene is not a linear structure that corresponds to a single product, but a modular structure that can be used in different ways to make different products.

As we will show in greater detail in Chapter 4, the sequences in the genome that act as molecular genes need not be physically distinct: they can overlap or occur inside one another (in the same direction on the DNA molecule or in reverse on the opposite strand). The relationship between structural genes and gene-like functions is not one-to-one but many-to-many: some gene products are made from more than one structural gene and more than one original transcript, and individual structural genes and their premature transcripts are processed into multiple products. Finally, the sequence of elements in the gene product depends on much more than the sequence of nucleotides in the structural gene: different sequence elements can be repeated, scrambled, and reversed in the product and the precise sequence of a gene product reflects not just the original DNA sequence but also its post-transcriptional and translational processing. Chapter 4, where we will describe some of these mechanisms with the help of examples, will put flesh on these bones and explore their philosophical implications.

3.5 A triumph of reductionism?

In the previous section we summarised some of the key discoveries of molecular genetics. These have been an impressive example of the power of reductionistic research. A huge amount has been learnt about the development and functioning of living organisms by understanding the interactions between the molecules of which they are composed, and no one can doubt that a great deal more will be revealed in the near future. The first philosophers to consider molecular genetics expected to find a successful reduction in two senses: the reduction of an older theory to a new and superior theory and the reduction of higher-level phenomena to lower-level phenomena (Schaffner 1969; Ruse 1971). But explaining in detail how one theory (or domain of phenomena) reduces to the other proved more difficult than expected and many philosophers of biology, starting with David Hull (1972), have been anti-reductionists. In the 1980s, partly due to the influential work of Philip Kitcher (1984), anti-reductionism became close to a consensus. It may seem paradoxical to find an anti-reductionist consensus about such an obvious triumph for reductionistic research, but philosophical disputes about reduction and molecular biology are primarily disputes about specific models of reduction. Most anti-reductionists allow for a fair amount of what in common-sense terms would be called reduction, and are 'anti-reductionists' only when contrasted to stronger forms of reductionism.

Many different senses of 'reduction' have been in play in this debate and we will need to distinguish some of them here before we can address which, if any, apply to the relationship between molecular biology and other fields. One sense of reductionism is *methodological reductionism*. This states that the most fruitful investigative strategy is the decomposition of systems into their component parts. The successes of molecular biology described in this chapter certainly conform to this prescription. But this does not mean that molecular biology will remain methodologically reductionist. Some of the creators of molecular biology, such as phage-group founder Max Delbruck, saw the reductionist phase of research as only the first of several phases of research needed to understand living systems (Sloan and Fogel 2012). Today it is common for leading biological researchers to argue that a more integrative approach is needed if the extraordinary amount of knowledge now available about living systems at the molecular level is to add up to an actual understanding of how those systems work (e.g. Noble 2006). Efforts to produce such an integrated understanding are often pursued under the heading of 'systems biology'. In the next two chapters we will encounter two themes in more recent molecular biology that run counter to methodological reductionism. In Chapter 4 we describe how the activity of molecular components is regulated by the other components with which they interact. In Chapter 5 we describe how the activity of the genome as a whole depends on the broader context in which it is situated.

A very different kind of reductionism is metaphysical or *ontological reductionism*. This is the idea that living systems are exhaustively composed of physical components. Ontological reductionism is very broadly accepted both in philosophy and biology, albeit with very little consensus regarding its implications for reduction in other senses. Alexander Rosenberg (1978) introduced into the philosophy of biology a related idea, *supervenience*, which clarifies the implications of ontological reductionism: biological phenomena supervene on physical phenomena in the sense that the biological phenomena cannot change without an accompanying change in the physical phenomena. One very weak form of reductionism, 'token-token reductionism', amounts to no more than this claim of supervenience of the biological on the physical. It says that each individual instance (token) of a biological phenomenon is identical to an individual instance (token) of some physical phenomena. For example, every Mendelian allele corresponds to some sequence of DNA. But this says nothing about how those *types* of phenomena are related. As we will see below, while some Mendelian alleles correspond to sequences that are molecular genes, others do not.

The philosophical literature on reductionism in biology is primarily concerned with *epistemic reductionism*. Epistemic reductionism is a family of claims about the relationship between different scientific domains and their bodies of knowledge. The classical form of epistemic reduction is the reduction of one theory to another. The early debates about reduction in philosophy of biology concerned whether the concepts and theories of classical genetics could be reduced to the concepts and theories of molecular genetics. These questions were initially framed in the light of Ernst Nagel's model of reduction, in which the theory to be reduced is deduced from the reducing theory with the help of coordinating definitions, or bridge principles, which relate the vocabulary of the two theories (Nagel 1961; Schaffner 1967; Schaffner 1969). A powerful critique of this approach was soon forthcoming from David L. Hull, who argued that neither classical nor molecular genetics contain exceptionless scientific laws that embody the content of the theory and can be lined up on either side of a formal deduction (Hull 1972, 1974, 1975). We have already touched on this point at the end of Chapter 2.

Hull also introduced a version of the 'multiple realisability' argument which was used to argue against reductionism in many different sciences in the 1970s and 1980s (Fodor 1974; Rosenberg 1978; Kitcher 1984; for a critique, see Sober 1999). Hull showed that many phenomena described in classical genetics can be instantiated by a range of different molecular mechanisms, which is an obstacle to defining those classical genetic phenomena in molecular terms. A case in point is the phenomenon of dominance. In classical genetics dominance means that the phenotype of the heterozygote, with one copy of the dominant allele, is the same as the phenotype of the homozygote with two copies of that allele (see Chapter 2). Dominance is often explained by the fact that the recessive allele is a mutation which impairs the function of the molecular gene: if one copy of the gene can do the job of two, then this will produce the phenomenon of dominance. But some loss-of-function mutations are not recessive, and several other mechanisms can produce the phenomenon of dominance. To this day no molecular account of dominance has captured the concept as it figured in classical genetics (Sarkar 2005; Falk 2009). Nor, argued Hull (1972), does the progress of biology require a molecular definition of this Mendelian phenomenon.

When read in the light of the Nagel model of theory reduction, Hull's arguments seemed to leave only two alternatives: anti-reductionism, which as we have noted seems paradoxical given the obvious successes of molecular genetics, or eliminitivism, the view that Mendelian genetics was refuted and replaced by molecular biology. One response to this dilemma was Kenneth Schaffner's generalised reduction-replacement (GRR) model which allowed the traditional, deductive account of theory reduction to acknowledge a richer set of alternatives than the stark dichotomy of reduction or elimination (summarised in Schaffner 1993). However, Schaffner himself conceded that the deductive approach to theory reduction did little to explain what had happened in the development of biology. Schaffner and others began to develop more realistic models of theory structure in genetics and molecular biology, and to rethink reduction using accounts of explanation other than logical deduction (Darden and Maull 1977; Schaffner 1993; Schaffner 1996).

Since the 1990s the emphasis in philosophy of biology has shifted from applying models of reduction to biology to using cases of successful biological research to develop more adequate models of reduction:

> The manifold difficulties encountered in applying Nagelian theory reduction [...] seemed to fit a problematic pattern – misconstruing biological reasoning with philosophical accounts of science forged on physical science examples. Philosophical issues in biology seemed to require distinct analyses that are more sensitive to empirical research in biology. (Brigandt and Love 2008)

One element of this shift is the recognition that theories may not be the relevant units of analysis. We argued in Chapter 2 that the primary achievement of classical genetics was not the theory of heredity embodied in Mendel's two laws and their later refinements (Falk 2004; Waters 2004; Falk 2007). Classical genetics was an investigative practice in which geneticists could explore broader biological questions. The genetic analysis of particular phenotypes provided data bearing on a wide range of biological questions, such as the basis of sexual differentiation. The classical molecular gene concept was the product of a highly successful attempt to identify the physical basis of the Mendelian gene, a research effort which itself employed the tools of genetic analysis. In the light of this, the question of epistemic reduction can be reconstrued as whether Mendelian genetic analysis has been reduced to some sophisticated form of analysis based on molecular genetics.

We would argue that this has not occurred. Although the material identity of the gene became dominant with the promulgation of the classical molecular gene concept, it could not entirely displace the other identity of the gene as an instrument for genetic analysis. As long as the patterns of experimental reasoning that make up genetic analysis are used, it is necessary to think of genes as Mendelian alleles. Marcel Weber's insightful comparison of Mendelian and molecular analyses of *Drosophila* loci concludes that 'even though the classical gene concept had long been abandoned at the theoretical level, it continues to function in experimental practice up to the present' (Weber 2005, 223), and Falk has argued that the patterns of reasoning distinctive of classical genetic analysis are still to be applied when what is being hybridised are not whole organisms but two pieces of DNA strands in vitro (Falk 2009).

This observation would be a mere quibble if the pieces of DNA picked out by the instrumental, Mendelian conception of the gene were always sequences that are also genes according to the classical molecular gene concept. But this is not the case. The definitive function of the molecular gene is to act as a linear template for the synthesis of biomolecules. But there are now known to be many other ways in which DNA sequences can play a role in the development of phenotypes. When one of these other pieces of DNA comes in two or more forms with different phenotypic effects, they will behave as Mendelian alleles and they can be investigated via genetic analysis. Even if they are not called genes, they are implicitly treated as such. However, such is the flexibility of scientific language that often they *are* called genes, but only when speaking in an appropriate context. For example, when a medical geneticist is seeking the 'genes for' a disorder she is looking for Mendelian alleles, sections of chromosome whose inheritance explains the phenotypic differences observed in patients. Translated into molecular terms these sequences may turn out not to be molecular genes, sequences that act as templates for the synthesis of a biomolecule and the immediate regulatory apparatus of such sequences.

A particularly clear example of the interaction of the Mendelian and molecular identities of the gene comes from studies of the gene Lmbr1 in the mouse and its homologue on human chromosome seven (Lettice et al. 2002). This locus is known to house an allele which produces abnormal limb development in both mice and humans. But further analysis of that locus shows that the gene in which the mutation is located plays no role in the development of these abnormalities. Instead, embedded in that gene is a sequence which acts to regulate the gene 'sonic hedgehog' (shh), located about one million DNA nucleotides away on the same chromosome, a gene known to be involved in the relevant aspects of limb development. The regulatory element at the original locus is not a molecular gene; on Waters' (1994) analysis it is not even part of the molecular gene which is found at that locus, since the mutations in question do not affect the products of that gene. But this regulatory element is the Mendelian allele for this kind of abnormal limb development. Conversely, shh is a paradigmatic molecular gene, but there is no allele of shh which is the Mendelian allele for this kind of abnormal limb development. Instead, in one experimental context, that of hunting for the mutation responsible for the phenotype, the gene takes on its Mendelian identity, while in the other context, that of analysing the sequence, the gene takes on its molecular identity. In many cases these two identities converge on the same sequence of DNA, but sometimes they do not.

So one clear sense in which Mendelian genetics does not reduce to molecular genetics is that it is not superseded by molecular genetics but remains alongside it as another way of thinking about DNA. Molecular genetics did not reduce *or* replace Mendelian genetics, but enriched genetics with another way of thinking about genes. As a result, biologists today have two different ways of thinking about genes: as Mendelian alleles and as sequences that template for a product. They move smoothly between these two contextually activated representations of the gene as they move from one kind of research to another.

3.5.1 Explanatory reductionism and mechanistic explanation

Attempts to show that the theory of classical Mendelian genetics can or cannot be deduced from a new theory called 'molecular genetics' turned out to have very little to do with either genetics as scientific practice, or the way in which molecular biology is self-evidently a successful example of reductionistic research. The most active discussion of reduction today focuses on a very different form of epistemic reduction – 'explanatory reduction'. Explanatory reductionism is the idea that higher-level phenomena can be explained by lower-level phenomena and their interactions (Sarkar 2005, 67; Rosenberg 2006).

In contrast to the consensus against traditional Nagelian theory reduction, several major figures in philosophy of biology have recently come out in favour of explanatory reduction. According to Sahotra Sarkar (2005) and Alexander Rosenberg (2006) the success of molecular biology proves that biology can be *explained* by chemistry and physics. Weber (2005) has advanced a similar view, but explicitly excludes evolutionary biology from this form of reduction. This argument depends on the idea that an explanation of the behaviour of a system in terms of the organisation and interaction of its parts counts as a reductive explanation. But it can be questioned whether this is reductionism as traditionally conceived. Earlier authors contrasted explanations in which the organisation and interaction of the parts are critical with explanations that need only the simplest kinds of relationships between parts, and regarded only the second kind as truly reductionist (Wimsatt 1986a). For example, statistical thermodynamics explains the phenomena of pressure, volume, and temperature while treating the gas as an ensemble of identical molecules interacting through random collisions. It can be argued that taking complex interactions into account moves beyond a purely reductionist approach, because what explains the behaviour of the system is how the parts are arranged, and the constraints this imposes on the activity of each part, as much as the properties of the parts themselves. Hence the explanation is not produced only by the reductionistic strategy of identifying and characterising the parts, but equally by the integrative strategy of showing how the parts fit together into one very specific kind of whole.

In the explanations outlined earlier in this chapter, and in many of those in subsequent chapters, the interactions between component parts of the

system are not determined solely by their own and their interacting part-
ners' molecular structure, but by the overall organisation in which they are
embedded:

> It can be argued that the impressive progress of the most reductionist of the
> biological sciences, molecular biology, is in fact helping to reinforce a
> scenario proposed long ago by the holist camp – the essence of a biological
> system is in the emergent properties of its interacting component parts.
> (Schlichting and Pigliucci 1998, 254; see also Gilbert and Sarkar 2000)

Philosophers Carl Craver and William Bechtel have argued that in such explanations we see an unproblematic, mechanistic version of the traditional anti-
reductionist idea of 'top-down causation', or, in their words, 'mechanistically
mediated effects'. The whole causes (by constraining) the behaviour of the
parts (Craver and Bechtel 2007).

What counts as a 'reductive' explanation is partly stipulative, but the
substantive point is that molecular biology has features that would traditionally have been regarded as anti-reductionist, as well as reductionist features. The explanations offered by molecular biology are paradigmatic mechanistic explanations, whose non-reductionist features Bechtel summarises as
follows:

> Mechanistic explanations also recognize the *fundamental role of organization* in
> enabling mechanisms to engage their environments as units (as well as the
> role of yet higher-level structures in constraining such engagement [. . .] It is
> typically the *higher-level disciplines that have the tools for discovering the organization
> within and between mechanisms*. Although these inquiries are constrained by the
> knowledge of the parts and operations constituting the mechanism, they
> make their own autonomous contribution to understanding how a
> mechanism actually behaves. Thus, mechanistic explanations provide a
> strong sense of autonomy for higher levels of organization and the inquiries
> addressing them even while recognizing the *distinctive contributions of
> reductionistic research investigating the operations of the lower level components*.
> (Bechtel 2007, quoted from abstract of online document; italics added)

The features of mechanistic explanation which Bechtel emphasises in this
passage will be amply exemplified in Chapter 4 and Chapter 5.

3.6 Conclusion

We have seen that the birth of molecular biology did not replace the practice of genetic analysis found in classical genetics, but rather added a molecular dimension to genetic analysis. More importantly the birth of molecular biology introduced an entirely new research area whose focus was the structural analysis of genes, the elucidation of their functional relationship to their products, and the regulation of their expression. The discovery of the molecular structure of the gene turned the hypothetical material gene into a well-defined molecular structure. In addition it redefined its relationship to the phenotype, and so the classical molecular gene acquired a new function – the production of a gene product – as well as a definite structure. This added a new research focus to the original focus of classical genetics. While classical genetics was a theory of heredity, molecular genetics was a theory of heredity and development. The contribution of genes to development was no longer black-boxed, but opened up by the unravelling of protein synthesis, the regulation of gene expression in the cell, and ultimately the development and differentiation of the organism.

Attempts to understand these developments by asking whether molecular genetics reduced or replaced Mendelian genetics were not successful. Genetics cannot be fitted into the simple model of growing knowledge about an object called 'the gene'. Instead, various uses of the term 'gene' relate to different kinds of research, and different representations of the gene that feature in those forms of research. Each of these ways of thinking is grounded in real facts about what the underlying molecules are doing, but it is not possible to reduce them all to one uniform way of chopping nucleic acid sequences up into 'genes'. The Mendelian gene is alive and well alongside its molecular descendant.

Finally, we have begun to argue here that the success of reductionistic research strategies in molecular biology does not vindicate a wholesale explanatory reductionism. This argument will be developed further in the following two chapters.

Further reading

The history of molecular genetics has been well served by historians of science. Two fine examples are Robert Olby's *The Path to the Double Helix* (1974) and Michel

Morange's *A History of Molecular Biology* (2000 [1998]). A good addition to these two would be Robert Olby's intellectual biography *Francis Crick: Hunter of Life's Secrets* (2009). The complex topic of reductionism in biology is ably summarised in an encyclopedia article by Ingo Brigandt and Alan Love (2008). Sahotra Sarkar's *Genetics and Reductionism* (1998) is a careful, book-length exploration of the topic. Two important works on mechanistic explanation are William Bechtel's *Discovering Cell Mechanisms* (2006) and Carl Craver's *Explaining the Brain* (2009).

4   The reactive genome

> The particulate gene has shaped thinking in the biological sciences over
> the past century. But attempts to translate such a complex concept into
> a discrete physical structure with clearly defined boundaries were
> always likely to be problematic, and now seem doomed to failure.
> Instead, the gene has become a flexible entity with borders that are
> defined by a combination of spatial organization and location, the
> ability to respond specifically to a particular set of cellular signals, and
> the relationship between expression patterns and the final phenotypic
> effect.
>
> (Dillon 2003, 457)

4.1 Postgenomics

This chapter explores the dramatic changes that have occurred in the understanding of genomes and their components in the 'postgenomic era' of molecular biology – the period that followed the publication of the draft human
genome sequence in 2001. After outlining some of those changes we go on in
4.2 to examine their implications for the gene. We argue that the molecular
conception of the gene introduced in the last chapter has evolved further. A
relatively conservative conception of the gene as a structural unit, only slightly
more complex than that envisaged by molecular biologists in the 1970s, continues to play a role in research. But this conception no longer fulfills the
central role of the molecular gene concept, which is to facilitate the study of
how the specificity of biomolecules is related to the informational specificity
of DNA sequences. For this, a more flexible conception of the gene is required,
as biological commentators have noticed (Gerstein et al. 2007). We refer to
this new conception as the 'postgenomic gene' (Griffiths and Stotz 2006) and

to the more conservative structural conception as the 'nominal gene' (Burian 2004).

The idea of informational specificity, or Crick information, introduced in Chapter 3 is central to the arguments presented in this chapter. In 4.3 we consider a recent philosophical reinterpretation of Crick's 'Central Dogma'. This is the claim that DNA sequences are the only 'causally specific actual difference maker' with respect to the specificity of gene products (Weber 2006; Waters 2007; see also Woodward 2010). These philosophers use the well-known theory of causation due to James Woodward (2003) to argue that whatever processes may be needed to regulate the expression of the genome, the source of biological specificity – the Crick information – is always the DNA sequence itself. In 4.4 and 4.5 we show that this argument ignores key aspects of how the postgenomic gene makes its products. We describe cellular processes which act as causally specific actual difference makers through the activation, selection, and creation of Crick information. In 4.6 we introduce the concepts of 'genetic underdetermination and amplification', 'distributed specificity', and 'molecular epigenesis' to better capture the actual causal relationships between the coding sequence, sequence-modifying factors, and gene products. Finally, in 4.7, we introduce 'systems biology', the name given by biologists to efforts to study the interaction and integration of the component parts discovered by molecular biology.

4.1.1 What is the 'postgenomic era?

The sequencing of the human genome led to the streamlining of sequencing techniques and hence an explosion of sequencing projects for other genomes. As of 1 February 2011 the number of published genome sequences had reached 1,580, not including about 306 'metagenomes' (the genetic material of a whole ecosystem, such as the human gut metagenome). A further 9,400 genome projects were ongoing.[1] This flood of data, and the bioinformatics tools developed to deal with it, has made possible for the first time the systematic exploration of the contents of genomes. The findings of the postgenomic era include surprising ways in which DNA performs and regulates its traditional protein-coding function, and a myriad of new non-coding functions for DNA.

---

[1] Genome OnLine Database (GOLD): www.genomesonline.org/

These new roles for the genome include the production of a range of previously unknown RNA products with enzymatic, structural, and regulatory functions.

The final draft of the human genome annotated just 1.5 per cent of the sequenced DNA as protein-coding genes, 25 per cent as introns, an unknown amount as regulatory sequences, and more than 50 per cent as transposable elements and pseudogenes, leading the scientific community to question both the previous emphasis on protein-coding genes and the use of the derogatory term 'junk DNA' for other regions of the genome. Another reason for interest in the non-coding regions of the genome was the 'N-value' or 'G-value' paradox (for 'number' or 'gene'), according to which the number of protein-coding sequences in a genome is not closely related to the complexity of the organism (Claverie 2001; Harrison et al. 2002). However, the ratio of transcribed, non-coding DNA to coding DNA shows a positive correlation with the organism's complexity, as measured by the size of its proteome (the set of proteins made by the organism) (Mattick 2001; Gagen and Mattick 2004). Early investigations revealed a large number of previously unannotated transcripts originating from within introns, from regions outside known genes, and from the complementary (antisense) strand of known genes. These previously unexplored regions were termed the 'dark matter of the genome' (Johnson et al. 2005) and in 2010 *Nature* listed 'shining a light on the genome's "dark matter"' as one of the 'insights of the decade' (Pennisi 2010). While many of these newly discovered transcripts show little protein-coding capacity, others extend the physical boundaries and genomic organisation of annotated genic regions. Transcripts can overlap coding regions in splice variants with novel 3' or 5' ends that can at times extend hundreds of kilobases. The idea of a distinct molecular gene with clearly defined boundaries has turned out to be overly optimistic, representing only a fraction of the coding capacities of the genome.

Several lines of enquiry suggest that the non-coding DNA which makes up most of eukaryote genomes serves some function. First, the comparison of the human genome sequences with others, notably the mouse, pointed to many evolutionary conserved non-coding sequences. Assuming that evolution would not conserve genuine 'junk' against mutation, this suggests that these sequences serve other functions (Hardison 2000; Shabalina et al. 2001). Second, as stated above, soon after the publication of the draft of the human genome several researchers pointed out that more than half of the DNA is transcribed into transcripts of unknown function (TUFs), more often than not without coding capacities (Chen et al. 2002). Third, biologists have found many new types of RNAs in addition to messenger, transfer, ribosomal, and small nuclear RNAs (the last of which functions to assemble the spliceosome; see Figure 4.2). With new types of RNA revealed on a regular basis, it has become challenging to keep track. For many of these RNAs at least some of their functions are now known: the majority seem to play roles in the regulation of protein production (Cawley et al. 2004; Mattick 2001; Mattick 2003; Mattick 2004). We will postpone discussion of non-coding RNAs to the next chapter, where we address their role in epigenetics.

In 2003 a National Institutes of Health project called the Encyclopedia of DNA Elements (ENCODE) set out to identify all the functional elements in the human genome, including transcriptional and other regulatory elements, gene and exon variants, alternative promoters in tissue-specific gene expression, and conserved non-coding elements (Oliver and Leblanc 2003; ENCODE Project Consortium 2007). At the beginning of September 2012 *Nature*, *Genome Research* and *Genome Biology* published the results from the ENCODE consortium in the form of thirty research papers (www.nature.com/encode; http://genome.cshlp.org/content/22/9.toc; http://genomebiology.com/series/ENCODE). Most notably, more than 80 per cent of the human genome was assigned at least one biochemical function, such as protein binding, transcribed RNA sequences, RNA binding, chromatin structure, DNA methylation, or histone modification (Maher 2012; ENCODE Project Consortium 2012).

We saw in Chapter 3 that the idea of informational specificity entered molecular biology in the 1950s to explain how molecular genes determine the biological specificity of gene products (Crick 1958). The idea of 'genetic programmes', to be discussed in more detail in Chapter 6, arrived in biology at around the same time, with the assumption that the programme would be written in DNA just as contemporary computer programs were written on magnetic tape (Mayr 1961). So the expectation was that the specificity of gene products would correspond in a fairly straightforward way to the informational specificity in coding regions of DNA. Another discovery of the postgenomic era has been the discrepancy between the number of genes in a genome and the number of products derived from them. For example, the human proteome outnumbers the number of discrete protein-coding genes by at least one order of magnitude. The human genome contains in the region of 20–25,000 genes (the correct number is still not known), while predictions have given numbers as high as 1 million proteins (Mueller et al. 2007). As we will show at length in 4.4 and 4.5, this discrepancy is explained by the fact that cellular mechanisms use the same coding region to make many different products, and combine resources from different coding regions to make products. The discovery that many other factors help to determine which products are made from the coding regions in the genome naturally led to these other factors being described in the same informational language as the coding regions themselves. For example, biologists began to talk about a 'histone code' to describe the way in which modifications to the histone molecules around which DNA is wound in the chromosome exercise a regulatory function. Different combinations of histone modifications can switch particular stretches of DNA from transcriptionally active to transcriptionally silent states. The range of RNA products involved in gene expression has also been described as a secondary code (Gibbs 2003). The regulatory modules of the genome that bind transcription factors, activators, and repressors, regions known respectively as promoters, cis-regulatory modules, enhancers, and silencers, have been likened to subroutines, which combine with routines in the coding regions to form a huge operating system (Gerstein et al. 2007, 671). Last but not least, the assembly of splicing factors in a given cell at a given time has been compared to a cellular splice code (Liu and Elliott 2010). This 'code' consists of combinations of regulatory elements in pre-mRNA substrates and around two hundred RNA-binding proteins whose comparative ratios have a decisive influence on which splice variants will be produced. In most of these new ways of speaking the term 'information' is loosely based on Crick's use of the word information to refer to the causes of sequence in gene products: 'Information means here the precise determination of sequence, either of bases in the nucleic acid or of amino acid residues in the protein' (Crick 1958, 153). We will return to this new use of Crick information in 4.3.

The 'postgenomic era' in molecular biology in part means a change of focus from the gathering and archiving of genomic data to their analysis and use in the discovery of genome structure and function. But it also refers to the way in which the discovery that genomes make a far wider range of products (the transcriptome) than they have genes has redirected attention to the regulatory architecture that *uses* the sequence information in the genome. Rather than looking for causes in DNA sequence information, the focus has shifted towards how sequences are used in a transient and flexible way by the varied mechanisms which control gene expression. These mechanisms involve *regulated recruitment and combinatorial control* (see 3.3.3). Regulatory molecules are recruited by intra-cellular, inter-cellular, and extra-cellular environmental signals. These RNAs and proteins then form complexes and networks, allowing fine-grained combinatorial control of gene expression. The study of these complex mechanisms necessitates the use of computer modelling techniques, and the development and testing of those models requires sophisticated computational techniques for accessing and analysing large quantities of sequence and other molecular data. This has undoubtedly changed the outlook of biological research. While the molecular decades behind us were characterised by the attempt to decompose organisms into their smallest components, the postgenomic era has given rise to a 'systems-biological' outlook which seeks to reassemble these components to learn how they interact to form complex living systems. We describe some of the ideas and approaches that fall under the heading of 'systems biology' in 4.7.

4.2 The gene in the postgenomic era

Genome sequencing projects produce an avalanche of DNA sequence data that call for 'annotation', the interpretation of the sequence as a set of meaningful parts.

The developments described in the previous section mean that this is a far more complex task than simply looking for sequences which fit the model of the classical molecular gene described in 3.3. Such sequences certainly exist, but many transcripts do not correspond to these sequences in any straightforward way. The achievement of the classical molecular conception of the gene was to unite structure and function in a single unit: a gene is a sequence with a distinctive structure (promoter, open reading frame, adjacent regulatory region), which performs the definitive function of the gene: namely, providing a template for the production of a gene product. Subsequent developments have shown that genomes have many more complex ways to perform that function. Moreover, the information located at the DNA sequence level only partially determines which products are derived from it. So sequence annotation becomes a balancing act between applying knowledge of function to annotate structure from the 'top down' and using knowledge of structure to annotate from the 'bottom up'. The classical molecular definition of the gene leaves open many decisions about the boundaries of genes when annotating sequences, even when additional information, such as knowledge of transcripts derived from the sequence and similarity to sequences in other species, is taken into account. Practitioners stress that annotation is an open-ended process that depends on future evidence and subjective judgments:

> The goal of annotation is to map features on the genome, initially focusing on developing models for genes that encode proteins. Good annotation requires an assembled sequence and a repository of the evidence for important genome features such as transcripts and sequence homologies to known genes. The annotation itself adds critical and explanatory notes to the genome. Thus, annotation is an executive decision about the relevancy, accuracy, and quality of the evidence, and by definition exposes the curator's point of view. (Oliver and Leblanc 2003, 204.1)

The complexity of the task of gene annotation even before the full complexity of the genome became apparent can be seen in the definitions of the gene offered by the private and public consortia that carried out the Human Genome Project (Lander et al. 2001; Venter et al. 2001). Their definitions were designed to allow a definitive count of the number of genes in the light of the discovery of alternative splicing, the process in which the final mRNA transcript is processed by cutting out large non-coding sequences (*introns*) from the pre-mRNA and splicing together the remaining sequences (*exons*) in various combinations. Celera genomics defined a gene as 'a locus of co-transcribed exons' (Venter et al. 2001, 1317), and the Ensembl Gene Sweepstake Web page originally defined a gene as 'a set of connected transcripts' where 'connected' meant sharing one exon.[2] The latter definition implies that the transcripts from a single gene may share a set of exons but in such a way that no one exon is common to all of them (Gerstein et al. 2007, 671).

These definitions of a gene take account of the fact that one gene can make a range of different transcripts through alternative splicing. They partition the genome into regions of exons connected by the transcription process. But even these quite plastic definitions are stretched to breaking point by further complications. A canonical case of alternative splicing produces a range of proteins that are structurally related to one another because the proteins share functional modules coded for by the same exons. There exist other cases, however, where the products from a single locus in the genome are quite different from each other. The degree of difference between products depends on how many exons they share, but also on whether these shared exons are read in the

---

[2] http://web.archive.org/web/20050428090317/www.ensembl.org/Genesweep

same reading frame. It is the precise nucleotide at which reading begins that determines which codons a DNA sequence contains. Starting at a different nucleotide is called 'frameshift', and produces a string of completely different codons. Frameshift makes it possible to derive two completely different products from the same sequence. Other highly divergent products are produced when two transcripts from the same locus overlap only at one end. Or one transcript may be entirely contained within an intron of another! Finally, two adjacent genes are sometimes 'co-transcribed' so that when the resulting transcript is translated it produces a completely new 'fusion protein' sharing parts from both genes. Cases like these in which very different transcripts originate from the same locus are often described as overlapping but distinct genes, rather than as alternative splicing of the same gene. That stipulation helps to preserve something closer to a unique correspondence between a gene and class of closely related transcripts, but it should be recognised that it *is* a stipulation.

There are also cases of *trans*-splicing, where a final mRNA transcript is processed from two or more independently transcribed pre-mRNAs.[3] *Trans*-splicing allows the fusion of transcripts from genes which are located on different parts of a chromosome, or even on different chromosomes. It also allows multiple transcripts from the same sequence, or transcripts from both halves of the double helix ('sense' and 'antisense' strands, running in opposite directions) to be spliced together. This produces final transcripts which may contain multiple copies of the same exon, or exons whose order has been 'scrambled' with respect to their order in the original DNA sequence (for examples of many of these splicing phenomena, see Stotz et al. 2006).

In recognition of the difficulties in applying the classical molecular conception of the gene to such non-canonical cases, it has been suggested that biologists employ a kind of 'consensus gene', a stereotype combining features from a number of exemplary cases (Fogle 2000). The consensus gene is based on a collection of flexibly applied features of well-established genes. A stretch of DNA is considered to be a gene, if it has enough of these features – for example, it contains an open reading frame and a well-defined promoter sequence such as a TATA box, and is transcribed into an RNA molecule which is

---

[3] In this context the prefix *trans-* and its opposite *cis-* are used slightly differently from the uses encountered in Chapter 2 and Chapter 3. *Cis*-splicing means splicing exons from a single pre-mRNA. *Trans*-splicing means using exons from more than one pre-mRNA, even if those pre-mRNAs are from the same chromosome.

processed into a polyadenylated transcript. The originator of this view, Thomas Fogle (2000), has argued that by combining structural and functional features into a single stereotype, the consensus gene hides both the diversity of DNA sequences that can perform the same function and the diversity of functions that DNA sequences perform. In other words, the consensus gene inherently distracts attention from problematic cases.

Working at the cutting edge of contemporary genomics can induce an extremely deflationary view of the gene. Some molecular biologists, realising that the concept of 'gene transcription' may not suffice to capture the variation in expressed genomic sequences, have proposed the more general term of 'genome transcription' to allow for the incorporation of transcripts from outside the border of canonical genes and the production of non-coding RNAs of all shapes and sizes with a multitude of functions. From this new perspective the classical molecular conception of the gene looks rather like 'statistical peaks within a wider pattern of genome expression' (Finta and Zaphiropoulos 2001, 160). One pragmatic, technological reason that today's biologists are prepared to consider such radical options is that the challenge of automated gene annotation has turned the apparently semantic issue of the definition of a gene into a pressing and practical one as the limitations of a purely structural, sequence-based definition of the gene have become apparent. An influential recent review concludes that 'one solution for annotating genes in sequenced genomes may be to return to the original definition of a gene – a sequence encoding a functional product – and use functional genomics to identify them' (Snyder and Gerstein 2003, 260). This is a top-down or functionalist approach to the gene, treating any sequence or set of sequences that can make a product as a gene, no matter how structurally complex they may be. Those who hold this view find their position strengthened by the results of the ENCODE project and the transcriptional and post-transcriptional processes it revealed. They pose the question 'What is a gene, post-ENCODE?' and answer:

> A gene is a union of genomic sequences encoding a coherent set of potentially overlapping functional products. Our definition sidesteps the complexities of regulation and transcription by removing the former altogether from the definition and arguing that final, functional gene products (rather than intermediate transcripts) should be used to group together entities associated with a single gene. It also manifests how integral the concept of biological function is in defining genes. (Gerstein et al. 2007)

By current definition the sequences involved in the *trans*-splicing cases described above would probably be counted as separate genes, while this newly proposed definition would treat the set of sequences as one gene because they cooperatively code for a functional product. Of course, they might still count as separate genes with respect to other products derived from them. The key word in the proposed new definition is 'functional', because that means providing evidence of the cellular process in which this newly detected protein or RNA is employed. Gerstein et al. proposed to throw overboard the long-held ideal that a molecular gene definition must combine functional and structural criteria. That is not very far from our own earlier suggestion that genes are ways in which cells utilise available template resources to create the biomolecules that are needed in a specific place at a specific time: genes are things an organism can do with its genome (Stotz et al. 2006)!

The paradox of the gene in postgenomic biology is that we are 'currently left with a rather abstract, open and generalized concept of the gene, even though our comprehension of the structure and organization of the genetic material has greatly increased' (Portin 1993, 173). Two decades later this is far more dramatically true than when the geneticist Petter Portin made that remark. The gene concept, however, plays a role in many other research contexts besides cutting-edge genomics. We have therefore suggested elsewhere that three complementary answers to the question 'What is a gene?' are needed to accurately depict the state of contemporary biology. The first of these is the traditional Mendelian gene, which is still needed today as we described in Chapter 3. The other two are descendants of the classical molecular gene, which we call the 'postgenomic gene' (Griffiths and Stotz 2006, 2007) and the 'nominal gene' (Burian 2004).

4.2.1 The postgenomic gene

We use this phrase to refer to the entities that continue to play the functional role of the molecular gene – making gene products – in postgenomic biology. A postgenomic gene is the collection of sequence elements that is the 'image' of the target molecule (the product whose activity we wish to understand) in the DNA, however fragmented or distorted that image may be. This comes close to Gerstein et al.'s (2007) suggestion of a purely functional gene concept. This conception of the gene retains a key feature of the classical molecular conception: namely, linear correspondence between gene and product. As we

argued in Chapter 3, linear correspondence plays a key role in the epistemology of the molecular bioscience because linear correspondence between
molecules is fundamental to biologists' ability to identify and manipulate
them. But although it is important to know the 'gene for' a molecule in this
sense, it does not matter whether that sequence or sequences is a gene in
the traditional sense. To put it less paradoxically, the utility of knowing the
DNA sequence(s) that underlie the production of the target molecule or its
precursors does not depend on whether it is possible to give a structural definition of a gene. Finding the underlying sequence(s) for a product remains
important even on the most deflationary, postgenomic view of what genes are
as structures in the genome.

4.2.2 The nominal gene

> The use of databases containing nucleotide sequences is well established.
> Codified as part of this process is a particular use of gene concepts on the basis
> of which one can identify various genes and count the number of genes in a
> given genome [. . .] I call genes, picked out in this way, nominal genes. A good
> way of parsing my argument is that nominal genes are a useful device for
> ensuring that our discourse is anchored in nucleotide sequences, but that
> nominal genes do not, and probably cannot, pick out all, only, or exactly the
> genes that are intended in many other parts of genetic work. (Burian 2004,
> 64–5)

It is hard to argue with Richard Burian that for many practical purposes genes
are simply sequences that have been annotated as genes and whose annotation
as such has been accepted by the scientific community. But, as Burian argues,
this does not imply that the scientific community has a clear understanding
of what makes a sequence a gene that needs only to be made explicit.

As we saw, gene annotation involves forming a judgment about the available evidence to arrive at a decision on how to label the genome sequence with
functional annotations like 'open reading frame', 'promoter', or 'pseudogene'.
Some underlying conception of the gene guides the decision, but this may be
no more than a stereotype of the kind Fogle (2000) describes as a 'consensus
gene'. Burian's (2004) point is that the practice of annotating some regions as
genes does not need to be underpinned by anything more substantial than
a stereotype in order to be useful. It is profoundly useful to have a database
of intelligent judgments as to which sequences have the kinds of features

that suggest they are doing the work of a gene. Some conservatism in this practice also has obvious value in promoting mutual understanding between researchers. But efforts like the ENCODE project create new knowledge about genome structure and function which may ultimately lead researchers to change their views. These new ideas about the gene can then be used to overhaul earlier annotations.

4.2.3 Gene-P and Gene-D

The philosopher and former cell biologist Lenny Moss has used the term 'Gene-P' for something very like the instrumental, Mendelian identity of the gene as we described it in Chapter 2 and Chapter 3 (Moss 2003). The P stands for 'phenotype', 'prediction', and 'preformation'. Gene-Ps are identified by their phenotypic effects, are used to predict the phenotypic results of hybridisation, and reflect what Moss calls 'instrumental preformationism' – a deliberate neglect of the ways in which the gene–phene relationship depends upon other factors. Moss contrasts Gene-P to what he calls 'Gene-D' (for 'development'). Gene-Ds are defined by their intrinsic capacity to template for gene products, and hence resemble the classical molecular gene as we characterised it in 3.3. The main difference between our approach and that of Moss is that Moss sees Gene-P and Gene-D as two fundamental ways of thinking about genetics which together provide a complete understanding of the gene. For example, where we distinguish the nominal gene and the postgenomic gene, Moss (personal communication) sees two versions of Gene-D corresponding to what are known as 'forward' and 'reverse' genetics. The postgenomic molecular gene embodies the traditional, 'forward' strategy of locating the template resources corresponding to a known phenotype. The nominal gene is a template resource whose use we set out to understand using the 'reverse' strategy. Moss describes the informational gene (see Chapter 6) as a conflation of Gene-P and Gene-D. We do not know how he would approach the 'abstract developmental genes' we introduce in Chapter 7.

However, even if we were to accept that many identities of the gene can be reduced to two fundamental ways of thinking about genes, and that those ways of thinking were specific to the gene rather than incarnations of some broader distinction such as that between structure and function, or between referential and attributive readings of the definite description 'the gene for x'

(Donnellan 1966), the detailed identities of the gene that emerge in particular research contexts would still be illuminating.

Both the postgenomic gene and the nominal gene are descendants of the classical molecular gene. The nominal gene tries to interpret sequences in a manner that stays as close to that original conception as the data will allow. An example of this is the decision to treat some regions from which multiple products are transcribed as cases of alternative splicing of a single gene, and others as distinct but overlapping genes. The postgenomic gene still has at its conceptual core the idea we stressed in our discussion of the molecular gene in Chapter 3. A gene is the source(s) in the DNA of the sequence specificity expressed in the product: it is where the 'Crick information' is to be found.

4.3 Distributed causal specificity

The postgenomic gene, like its classical molecular forebearer, is a source of sequence specificity for a linear gene product. C. Kenneth Waters maintains that it is the only real or important source of specificity for the gene product. It was Waters who first drew philosophical attention to the underlying rationale of the classical molecular conception of the gene. A molecular gene is a sequence of nucleotides which stands in a linear correspondence to the linear order of elements in a gene product – the nucleotides in an RNA or the amino acids in a polypeptide (Waters 1994, 2000). Waters recognised that this approach does not identify a single, determinate sequence of nucleotides as 'the gene' but rather allows the boundaries of each gene to change depending on which product the investigator has in mind, and which stage in the expression of that product. But although different products draw on different parts of the DNA sequence, in Waters' vision that sequence is the sole source of specificity.

In his recent work Waters has reiterated this view. He identifies the privileged role of the molecular gene in many biological explanations as that of an 'actual difference maker' with 'causal specificity' (Waters 2007, 572). While in essence he is rephrasing Crick's sequence hypothesis, he hopes to bolster his argument by showing that the privileged role of DNA with respect to other factors falls out of the application of James Woodward's widely accepted philosophical account of causation to genetics (Woodward 2003). His aim 'is to identify situations in which DNA is an ontologically distinctive cause, and to clarify the nature of its causal distinctiveness in these situations' (Waters 2007, 572).

We have no quarrel with the Woodwardian framework in which Waters chooses to analyse this issue. In many experimental situations biologists do aim to identify actual difference makers. However, as we will see in Chapter 7, Waters is wrong to assert that this is *always* what biologists seek to do. Sometimes they aim to identify as wide a range as possible of potential difference makers. We are also happy to accept Waters' direct challenge and show that DNA shares its sequence specificity with other cellular factors which also act as causally specific actual difference makers with respect to gene products. However, by combining a general discussion of how the Woodward framework might be applied in biology with this specific dispute Waters has muddied the waters. He chooses to analyse certain limited instances in which the causal processes that we have focused on in our earlier work – namely, the post-transcriptional regulation of gene expression and the role of the environment in behavioural development – are not present. Waters concludes that in his chosen cases DNA is the sole causally specific actual difference maker. He then accuses us of making a fundamental mistake about causal reasoning (Waters 2007, esp. fn. 5). In fact the differences between us have nothing to do with the nature of causation or causal reasoning, but turn entirely on which biological phenomena we choose to discuss. Waters' conflation of these two sources of disagreement has been quite influential. James Woodward (2010), for example, takes himself to be arguing against us in his very helpful presentation of how salient explanatory causes are distinguished for various biological phenomena. But this is only because he wrongly assumes that our argument for the importance of non-genetic causes is that all causes are equal. We will actually make use of Woodward's framework to *support* our position in Chapter 7.

Woodward's theory construes causation as a relationship between variables in a scientific representation of a system. There is a causal relationship between variables X and Y if it is possible to manipulate the value of Y by intervening to change the value of X. 'Intervention' here is a technical notion with various restrictions, so that, for example, changing a third variable Z that simultaneously changes X and Y does not count as intervening on X. Causal relationships between variables differ in how 'invariant' they are. Invariance is a measure of the range of values of X and Y, and of the other variables that characterise the system, across which the relationship between X and Y holds. But even relationships with very small invariance spaces are minimally causal.

Building on Woodward's account, Waters notes that Y may have many variables as potential causes, but if those variables do not actually vary, then their values will not explain any of the actual differences in the values of Y in some population. He therefore defines the notion of an 'actual difference maker' as follows:

> X is an actual difference maker with respect to Y in population p if and only if
>
> i. X causes Y (in Woodward's sense).
>
> ii. The value of Y actually varies among individuals in p.
>
> iii. The relationship X causes Y is invariant over at least parts of the space(s) of values that other variables actually take in p. (In other words, it is invariant with respect to a portion of the combinations of values the variables actually take in p.)
>
> iv. Actual variation in the value of X partially accounts for the actual variation of Y values in population p (via the relationship X causes Y).
>
> (Waters 2007, 571)

Finally, Waters notices that some variables only change the value of Y in a single way; for example, by always reducing Y to zero when X exceeds some threshold value. But others may be able to change the value of Y in very specific ways, depending on the precise value of X. He therefore introduces the notion of a *specific* actual difference maker – a variable X with a range of values that cause a range of values of Y. It should be noted that the distinction between specific and non-specific causes, developed more fully by Woodward (2010), is closely related to the distinction found in developmental biology between 'instructive' and 'permissive' causes (Gilbert 2000, 142–3).

In summary, a *specific actual difference maker* (we will abbreviate this as SAD) is something which differs across a range of actual cases, and which causes something else to differ in such a way that the specific differences in the effect depend on the specific differences in the cause.

With this apparatus in place, Waters sets out to identify the SADs for the nucleotide sequences of the RNA molecules found in a bacterial cell. He considers two candidates, the DNA sequences in the cell and RNA polymerase, the critical enzyme in transcription. He concludes that the DNA sequences are the SADs because only the DNA sequences will actually differ between one RNA molecule and another, while the polymerase, the critical enzyme in transcription, will be the same in each case. So the DNA sequences are the SADs for the RNA sequences. This is essentially a restatement of Crick's sequence hypothesis: only DNA is able to specify the linear order of the nucleotides in RNA (Crick 1958; Stotz 2006b). Turning to eukaryotic gene expression Waters acknowledges that post-transcriptional processes such as splicing change the sequence of the final product, hence he grants that splicing agents can be SADs too, but he minimises this concession by focusing on the pre-mRNA, the initial transcript before it has undergone any processing. He concludes that 'DNA is the causally specific actual difference maker with respect to the population of RNA molecules first synthesized in eukaryotic cells' (Waters 2007, 575).

What is puzzling about Waters' paper is that he takes his analysis of these carefully tailored cases to vindicate an explanatory emphasis on DNA relative to other causes quite generally. The authors he berates for suggesting that other factors can enjoy 'causal parity' with DNA are concerned with the relative importance of genetics, epigenetics and environment in development, and his arguments have no real bearing on this issue (in 7.3 we will see how a single-minded focus on SADs actually causes scientists to overlook important causes in behavioural development). Waters argues that the authors who emphasise the role of epigenetic and environment factors in development only do so because they believe that all causes are equal. This is 'fallacy of causal parity arguments' (2007, 572). 'Parity arguments,' says Waters, 'claim that picking out one cause, when in fact there are many, cannot be justified on ontological grounds because, after all, causes are causes' (553). But this is a parody. Parity arguments are actually about drawing the same kind of distinctions that Waters draws, and showing that in certain explanatory contexts factors other than genes are explanatorily relevant causes. Those of us who make these arguments have clearly and repeatedly denied that we advocate what Philip Kitcher (2001) christened 'causal democracy' (e.g. Griffiths and Knight 1998; Oyama 2000a; Griffiths and Gray 2005).

Not only do Waters' arguments fail to reach his intended targets, they fail to extend to most of contemporary molecular biology. In the sections that follow we will see that the role of acting as a SAD for sequences in gene products is not monopolised by DNA but is distributed among DNA sequences, regulatory RNAs, proteins, and environmental signals. We should also note that in eukaryotes pre-mRNAs are very short-lived intermediaries. While they are being transcribed at one end they are already being processed into their mature form at the other! This fact has now been conclusively confirmed for the entire human coding transcriptome by the ENCODE Project Consortium:

> [W]e confirm that splicing predominantly occurs during transcription [. . .] Cotranscriptional splicing provides an explanation for the increasing evidence connecting chromatin structure to splicing regulation, and we have observed that exons in the process of being spliced are enriched in a number of chromatin marks. (Djebali, Davis et al. 2012, 102)

These mature sequences are causally co-specified by *cis*-acting sequences interacting with a large range of *trans*-acting factors that carry sequence specificity for the particular slice sites and editing sites. These are the actual difference makers that *select* from and *modify* the pre-mRNA to produce the mature mRNA sequence. If we were to intervene in order to prevent the production of specific RNAs in a cell without interfering with all other possible splice variants from one coding sequence, then our natural intervention point would be the cellular splice or editing code, rather than the original DNA sequence. The results from the ENCODE project confirm that

> it is currently close to impossible to predict from the analysis of mammalian primary RNA sequence alone neither the entire exon–intron structure of transcripts nor their tissue specific expression pattern [. . .] It appears thus that other factors, not necessarily encoded in the sequence of the primary transcript, may play a role in splicing definition. (Tilgner et al. 2012, 1616)

Waters admits that splicing factors may share specificity with DNA sequences, but he tries to minimise the role of splicing factors in gene expression. He remarks in another paper that '[i]f differential RNA splicing occurs within the same cell structure at the same time, then differences in the linear sequences among these polypeptides [. . .] could be said to be caused by differences in splicing factors, rather than differences in DNA. It would still technically be true that different "split genes" were involved' (Waters 2006, 208). This passage contains two moves to downplay the way in which splicing factors provide specificity. First, Waters focuses on explaining differences between RNAs in a single cell at a single time rather than the RNAs in a cell at different times, or in different cells with the same genome. Assuming that each cell consistently produces only one splice variant at a time, this restriction would allow him to relegate splicing and other cellular factors to mere background conditions. However, in most cases of alternative splicing, splicing factors influence the *ratio* of splice variants in a particular cell at a time rather than selectively producing only certain splice variants: 'For most alternatively spliced transcripts there is no "default" or unregulated state; instead, the ratio of alternative splice forms observed for a given pre-mRNA results from a balance between positive and negative regulation' (Ladd and Cooper 2002, 3). This is indeed another fact confirmed by the results from the ENCODE project:

> First, isoform expression does not seem to follow a minimalistic strategy. Genes tend to express many isoforms simultaneously [. . .] Second, alternative isoforms within a gene are not expressed at similar levels, and one isoform dominates in a given condition—usually capturing a large fraction of the total gene expression [. . .] Third, about three-quarters of protein-coding genes have at least two different dominant/major isoforms [. . .] (Djebali, Davis et al. 2012, 103–4)

But Waters ultimately admits that alternative splice forms might occur in the same cell at the same time. His next move is to suggest that DNA sequences for different splice variants count as different 'split genes'. Unless the alternative products are very different from one another, this stipulation departs from conventional practice in molecular biology, where multiple isoforms of a protein are usually ascribed to one nominal gene as we described above. Presumably by calling his proposal 'technically true' Waters means that this stipulation would follow from his own account of the nature of the gene in which a gene is defined top-down by looking at the gene product and reading it back into the DNA. The only structural constraint on genes in his account is that they are made of DNA (Waters 1994). We are sympathetic to thinking about genes in this way when the relationship between gene and product is extremely complex. We introduced this idea in 4.2.1 with our 'postgenomic gene'. But in many cases of manageable complexity molecular biologists treat the underlying sequence as a nominal gene and allow this single, nominal gene to give rise to a range of alternative products. The nominal gene is defined by a loose family of both structural and functional constraints and represents an attempt to stay in touch with the classical molecular conception of the gene insofar as the data will allow this.

Overall, Waters' argument appears as an attempt to 'rescue' DNA as the (more or less) sole bearer of causal specificity. Waters' account downplays some of the major theoretical insights into genome structure and function revealed by contemporary molecular genetics and genomics, including surprising ways in which DNA performs its traditional gene-like functions, new un-gene-like functions, and other cellular structures that share some of DNA's cellular functions. His central claim, that the underlying DNA sequence is the main source of the biological specificity of gene products, with cases of shared specificity as rare exceptions, does not match current knowledge of genome structure and function.

The idea of specificity – first the stereo-chemical specificity of macromolecules and then the linear, informational specificity of nucleic acid molecules – has been the touchstone of molecular biology. It transformed our understanding of biological mechanism from a highly fluid and interactive process into an assembly of pieces each with its own specific and restricted part to play (Greenspan 2001). But the idea of the DNA sequence as the sole source of specificity does not seem to capture how complex organisms are regulated and organised. Comparing the human genome with its transcriptome reveals sequence information not encoded by the literal DNA code alone. Intra-cellular, inter-cellular, and extra-cellular environmental signals provide specificity via regulatory RNAs and proteins organised in expression mechanisms which have an impact on the final sequence of the gene product. The transcriptome is not specified by a limited number of distinct protein-coding genes but by the totality of what the cell can do with its genome. In the next two sections we will show how sequence information is *activated, selected,* and *created* by *causally specific* regulatory mechanisms of genome expression. The overall picture that emerges is one of *molecular epigenesis,* an idea we will expand on in 4.6.

4.4 The flexible genome: sequence activation and selection

A major theme in this chapter and in Chapter 3 has been the determination of the sequence of a gene product by the informational specificity of the underlying DNA sequence. We have also expressed this by saying that the DNA contains the 'Crick information' for the product. Our aim now is to show that other factors share the role of providing Crick information. In this section we describe mechanisms which differentially activate sequence information, so making a causally specific difference to which transcripts are produced in a particular cell at a particular time, and mechanisms which differentially select sequence information, thus making a causally specific difference to which sequence information is transcribed from a given portion of the genome. In 4.5 we describe mechanisms which create new sequence information, so making a causally specific difference to the linear order of elements in the gene product.

We do not deal here with the most fundamental kind of sequence selection, which is the modification of chromatin into a form in which the DNA is available for transcription. These processes are now often subsumed under the heading of epigenetics and we deal with them in Chapter 5.

4.4.1 Transcriptional regulation

The primary difference between the regulation of gene expression in prokaryotes and in eukaryotes is the elaborate regulatory architecture of eukaryotic genes which allows them to produce a greater number of gene products and to tailor their products to a wide variety of cellular conditions. The strategies employed to allow this fine-grained control are *combinatorial control*, meaning that a range of different causal factors can be brought together in different combinations, plus a high degree of integration between different regulatory signals, and the presence of alternative sets of regulators (Ptashne and Gann 2002, 115).

All gene expression mechanisms including transcriptional activation or inhibition, splicing, editing, and other co- and post-transcriptional processes, have in common the combinatorial interaction of multiple kinds of *cis*-acting sequence modules located upstream, downstream, and within the coding sequences. These bind a range of *trans*-acting proteins and RNAs. Many of these factors need to be transported from other locations in the cell or otherwise activated by intra- or extra-cellular signals, and recruited to join the transcriptional complex. One can say that *trans*-acting factors *function as mediators of environmental information to the genome*. Most *cis*- and *trans*-acting elements have in common that they are individually weak, not fully specific, and present in multiple copies. As we now know, 'there is little or no *constitutive* regulation in higher organisms; i.e., the differentiated state of normal cells is unstable and the environment regulates gene expression' (Bissell 1981, 27; quoted in Bissell 2003). Because of the structure of their complex, modular, but generally weak promoter sequences gene expression in eukaryotes always requires the recruitment of a large transcriptional machinery of *trans*-acting factors to the *cis*-acting modules through the action of specific transcription factors called 'activators'. These are one kind of *trans*-acting factor that bind to a special class of *cis*-acting sequences, known as enhancers; these can be a long distance

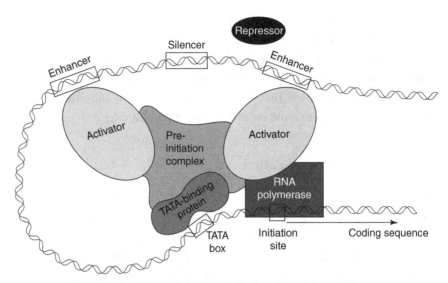

Figure 4.1 Simplified schema of the transcriptional machinery. Distant enhancers are shown with their activators (sometimes called specific transcription factors). These recruit the TATA box binding protein (and accessory protein, not depicted) and the pre-initiation complex, assembled out of a range of transcription factors and co-factors (details not shown). This complex recruits the RNA polymerase to the initiation (or transcriptional start) site. Distant silencers can bind repressors (shown in the background) that could down-regulate or even shut down transcription initiation. Not shown: the chromatin remodelling complex that renders the DNA open for transcription (see Figure 5.1), and *cis*-regulatory modules upstream of the promoter (TATA box) that bind further specific transcription factors.

from the coding sequence on the DNA strand, but are often in close physical proximity due to the three-dimensional structure into which the chromatin is folded, or can be folded with the help of proteins that can bend the DNA. The exact order and nature of the activator's recruitment is still largely unknown; however, we know that the full machinery comes in the form of separate, sometimes preassembled complexes made up of a large number of different proteins (see Figure 4.1): the activator complex assembles at the enhancer to recruit the chromatin remodelling complex (which renders the DNA accessible) and the TATA-binding proteins and associated factors, which bind to the TATA box or other protein-binding sequences of the promoter to recruit the holoenzyme (transcription enzyme RNA polymerase, specific transcription factors, and transcription co-factors). The key point to notice is that the factors influencing transcription can be combined in many different ways, assigning each single factor a slightly, but sometimes even a dramatically, different role. This *combinatorial control* allows the fine-tuning of gene expression in response to a range of incoming signals.

While activation is mostly described as a simple on-or-off decision, activation can also influence the selection of the actual sequence which is transcribed because of the common existence of alternative promoters and transcription start and termination sites. The differential use of promoters can also co-specify splice site selection when alternative first exons come with their own promoter (Dorn et al. 2001; Tasic et al. 2002). Moreover, while some components of the transcriptional machinery detach from the polymerase after successful activation, others will subsequently move along with the polymerase during the transcriptional process and will interact with the capping, splicing, polyadenylation, and editing machineries (Davidson 2001; Bentley 2002). They thus contribute to the sequence selection and creation processes described below.

As noted above, cells respond to intra- and extra-cellular signals such as hormonal or nutritional changes, with changes in gene expression. This is mediated through the environment-specific use of regulatory elements (Ptashne and Gann 2002). A common mechanism of activation is the phosphorylation of transcriptional regulators – the addition of a phosphate group to a site in the molecule which changes its conformational state and hence its stereochemical specificity. *After they are induced by environmental signalling factors, the specificity of many* trans-*acting factors is imposed through differential recruitment and combinatorial control.*

At least some of the mechanisms just described act as causally specific actual difference makers (SADs) because they make a difference to which product is produced from the same underlying DNA sequence. A cell can contain two transcripts from the same gene whose difference is caused by one of these transcriptional mechanisms. However, we see no reason to restrict the question of the sources of specificity to the production of gene products in one cell at one time, as Waters does. Equally relevant questions include why the same sequence of DNA produces different products at different times, or in different places. The mechanisms just reviewed make an important contribution to answering all these questions.

In the remainder of this section we describe the main mechanisms involved in the *selection* of sequences: namely, alternative splicing. Together with the mechanisms of activation and pre-selection they represent 'conservative' cases of distributed causal specificity because they leave intact the *linear order* of the original sequence in the gene product and only determine which parts of the sequence will be used. In 4.5 we will encounter more radical cases.

4.4.2 Alternative splicing

In eukaryotes, the DNA sequence is transcribed into a pre-messenger RNA from which the final mRNA transcript is made by cutting out the introns and splicing together the exons. At one time it was thought that alternative splicing only occurs in coding regions and exons were defined as sequences that are eventually translated to protein. It is now known that alternative splicing in regions that are not translated to protein is used to regulate the processing of the transcript, so exons are defined as sequences that are found in the mature, processed RNA. Moreover, many genes are now known to code for functional RNAs that are not translated into protein at all, and this new definition of exon can be applied to these genes too.

Biologists speak of alternative *cis*-splicing when more than one mature mRNA transcript results from the cutting and splicing of alternative exons from a single mRNA. If a coding sequence containing the four exons, 1234, were always to produce a final mRNA with three exons, there would be four possible alternative splice forms: 123, 124, 134, 234. Here one gene can produce four distinct products. Alternative splicing is not a rare phenomenon: current evidence suggests that a large majority of human genes ('nominal genes') undergo alternative splicing. *Agents other than the original coding sequence have to provide sufficient splice site specificity to control this diversification.*

Pre-mRNA splicing is the process by which two successive chemical reactions cleave the upstream exon from the intron and join ('ligate') it to the downstream exon. This takes place on the spliceosome, a dynamic complex of small nuclear RNAs and associated proteins (see Figure 4.2). This happens with every splicing process, so what specifies *alternative* splicing? The splice site sequences within the intron are generally small and weak and not always sufficient to specify alternative splicing. Splicing specificity must therefore be imposed through the assistance of additional *cis*-acting sequences located either in the adjoining exons or within the intron, and by *trans*-acting factors which bind to them. Either one or more specific splicing factors or the ratio of a range of common splicing factors can make the difference between

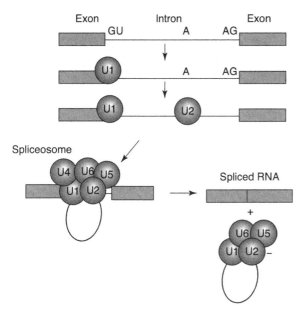

Figure 4.2 The assembly of the spliceosome complex. The schema depicts the five snRNPs (small nuclear ribonucleic proteins) that bind extrinsic serine/arginine-rich (SR) protein factors (not shown) and assemble in separate steps on the juxtaposed 5′ and 3′ splice sites and an anchor sequence not far from the splice site, all within the intron. The snRNAs that form the spliceosome complex are rich in uridine and are named U1, U2, U4, U5, and U6. They are involved in several RNA–RNA and RNA–protein interactions. The consensus splice site sequences in the intron are: 5′ splice site = GU (nucleotides), branch point or anchor sequence = A (nucleotide), 3′ splice site = AG (nucleotides).

the inclusion or exclusion of an exon. While exonic and intronic splicing enhancers (ESE and ISE) positively stimulate the spliceosome assembly at certain sites, exonic and intronic splicing silencers (ESS and ISS) block certain splicing choices (see Figure 4.3) (Smith and Valcarcel 2000). In other words, *the availability of certain* trans-*acting factors and the differential combinatorial binding of spliceosomal binding RNAs and proteins to splice sites and regulatory sequences (the 'cellular splice code') seem to be the major contributor to splicing specificity.*

Three major mechanisms are known that change the 'cellular code' for splice site selection: the synthesis of new splicing proteins by special regulator genes, the activation of splicing proteins through phosphorylation, and the movement of splicing regulatory proteins into the nucleus (Stamm 2002; Shin and Manley 2004). 'The combinatorial mechanism for the control of

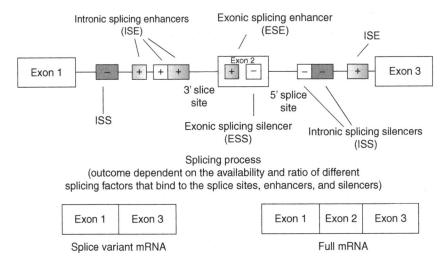

Figure 4.3 Schema of the distribution of *cis*-regulatory splicing modules for one exon. Besides the canonical splice sites (here very simplified) there exist a range of enhancers and inhibitors within the exon and in the two flanking introns (shaded boxes). Multiple copies of the same sequence (indicated by the same shade) bind the same splicing factors, either serine/arginine-rich (SR) splicing proteins or heterogeneous nuclear ribonucleoproteins (hnRNPs). The available ratio of splicing factors in the cell determines which splice variant will be produced.

alternative splicing [...] could allow cells to adjust splicing outcome (and consequently which proteins they express) rapidly in response to intra-cellular or extra-cellular cues, as well as contributing to the generation of protein diversity' (Bradbury 2005). In other words, *the cellular context imposes splice site specificity via the cellular splice code.*

4.4.3 Other sequence selection mechanisms

So far we have described how a single nominal gene produces different splice variants, but the genome also produces a variety of transcripts that are hard to attribute to a single nominal gene. Many transcripts contain exons from adjacent genes, or sequences from pseudogenes and intergenic regions which are 'co-transcribed' to produce a single pre-mRNA (Finta and Zaphiropoulos 2000a, 2002; Communi et al. 2001; Kapranov et al. 2005). Pseudogenes derive from partial gene duplications, which typically render the partial gene copies non-functional, hence their name. Nevertheless, it has been shown that some pseudogenes are transcribed and processed, and while their function (if any) is often unknown, in some cases their mRNA seems to exert a stabilising effect on the transcript of the homologous, functional gene from which they derive (Hirotsune et al. 2003).

Gene products may also be derived from so-called 'overlapping genes'. These include antisense transcripts, which are produced from a region of DNA which is already recognised as a gene, but are derived from the opposite strand of the double helix and therefore transcribed in the opposite direction. Another form of 'overlapping genes' involves transcribing almost the same sequence of DNA, but with slight differences meaning that the transcripts are read in an alternative reading frame and so give rise to completely distinct products (Blumenthal et al. 2002; Coelho et al. 2002). As noted above, the decision whether to regard such cases as overlapping genes or as radical instances of alternative splicing is usually made on the basis of whether the different gene products are closely related to one another and is to some extent conventional.

Instead of *mutually exclusive* alternative transcripts arising from the same DNA sequence, as is the case with alternative splicing and many overlapping genes, *multiple simultaneous* transcripts can occur from the same DNA sequence, based on a single pre-mRNA. This occurs when functional non-coding RNAs (such as microRNAs and snoRNAs) are derived from regions of a transcript which would be regarded as introns if we concentrated on the major product of that transcript. These RNAs may be involved in the regulation of the coding transcript of the same gene, but need not be.

In this section we have described the differential activation and differential selection of nucleotide sequences through a range of transcriptional and post-transcriptional mechanisms. These are 'conservative' examples of distributed specificity because in all these cases there remains a linear correspondence between the order of elements in the underlying DNA sequence and the product. The product may contain only selected fragments of the sequence information in the DNA sequence, but within each of those fragments the order of elements is determined by the corresponding DNA sequence, and the order of the fragments themselves is also determined by the order of the corresponding elements in the DNA. In 4.5 we will describe more 'radical' examples in which this is not the case. We assert, however, that even in the conservative cases, the regulatory mechanisms are sources of specificity.

It seems clear that Waters would deny this. First, he restricts his analysis to pre-mRNAs. But it is not reasonable to focus on pre-mRNAs in an account of the causes of the specificity of gene products. The whole point about specificity is that it is something possessed by functional gene products and which explains their functionality. Besides, the claim that pre-mRNAs are determined solely by the DNA sequence is close to trivial, as a pre-mRNA is defined as a transcript before it undergoes any processing. Because many of the 'post-transcriptional' processes described above actually happen simultaneously with transcription the pre-mRNA found in representations of transcription may never even exist as an actual molecule.

A second response from Waters would focus on the point we have conceded about the linear order of elements in these conservative cases. Waters could insist that sequence specificity or Crick information is restricted solely to factors that determine the linear order of elements. This would exclude from what needs to be explained all the features of products that we have shown depend on regulatory mechanisms, such as where the sequence begins and ends, and which elements occur in the sequence (as opposed to what order they occur in). Waters would have to exclude even the linear order of element itself whenever there is a boundary between parts of the sequences corresponding to different fragments of the DNA sequence! We can see no biological rationale for these restrictions. Crick did not enunciate the sequence hypothesis for its own sake. He enunciated it as an explanation of how cells confer biological specificity on their products. Consequently, it is more reasonable to read his phrase 'the precise determination of sequence' (Crick 1958, 153) to mean the determination of what the sequence is – where it begins, what it contains, and what order all those elements come in – rather than restricting it to the linear order of elements, and not even all of that.

4.5 The flexible genome: sequence creation

In the following 'radical' cases of sequence determination the *linear sequence* of the final product is not mirrored by the DNA sequence but is extensively *scrambled*, *modified* or literally *created* through a variety of co- and post-transcriptional mechanisms. Although we have separated them for reasons of exposition, these mechanisms are often interdependent with the mechanisms of sequence activation and selection.

4.5.1 *Trans*-splicing

Biologists speak of *trans*-splicing when a final mRNA transcript is processed from two or more independently transcribed pre-mRNAs. These separate pre-mRNAs can be derived from DNA sequences that are far apart in the genome, but they can also be multiple transcripts from the very same DNA sequence. The latter case allows the inclusion of multiple copies of the same exons. It also allows the original order of exons to be scrambled in the final transcript. To use the earlier example of a gene with four exons, 1234, that always produces a three-exon mRNA, *trans*-splicing allows additional variants such as 231 or 233. Some documented cases of *trans*-splicing involve two or more genes which originated through gene duplication, which have diverged from each other with respect to sequence and function, and which are now interchanging their exons (Finta and Zaphiropoulos 2000b). Recent studies have provided more evidence for the abundance of these so-called chimeric transcripts and their importance in humans.

> The biological and evolutionary importance of these chimeric transcripts is underscored by (1) the non-random interconnections of genes involved, (2) the greater phylogenetic depth of the genes involved in many chimeric interactions, (3) the coordination of the expression of connected genes and (4) the close in vivo and three dimensional proximity of the genomic regions being transcribed and contributing to parts of the chimeric RNAs. (Djebali, Lagarde et al. 2012, Abstract; see also Frenkel-Morgenstern et al. 2012)

Trans-splicing phenomena are inconsistent with Crick's sequence hypothesis because *they change in a regulated way the linear order of the elements in the product with respect to the order of the elements in the DNA from which those elements are derived.*

Mechanisms for splicing in *trans* are supported by splicing agents that seem to be split versions of their equivalent *cis*-splicing agents (Wissinger et al. 1991; Malek and Knoop 1998; Sturm and Campbell 1999; Caudevilla et al. 2001; Rivier et al. 2001). Finding mechanisms related to *cis*-splicing is not surprising: in genes with very long introns splicing happens almost in *trans* because the two ends with the relevant sequences are so far apart.

4.5.2 RNA editing

RNA editing is another prevalent mechanism of sequence modification that can significantly amplify the coding sequences in the genome to produce a much larger transcriptome (see Gott and Emeson 2000 for a good overview). RNA editing disturbs the linear correspondence between gene and product, in some cases to a very large extent. But while *trans*-splicing did so by *scrambling* the order of the primary DNA sequence, RNA editing *changes* the primary sequence of mRNA during or after its transcription via the site-specific insertion, deletion, or substitution of one of the four nucleotides, adenine (A), uracil (U) (which replaces thymine [T] in RNA), cytosine (C), and guanine (G) (Gray 2003). Editing of one kind or another has been described in almost all eukaryotes including humans, both in the nucleus and organelle genomes, but not yet in bacteria. Co-transcriptional editing happens at the pre-mRNA state, while post-transcriptional editing takes place in the final mRNA. RNA editing has been said to create 'cryptogenes' because it leads to gene products whose 'image in the DNA' is unrecognisable. RNA editing affects many kinds of RNA (messenger RNAs, transfer RNAs, ribosomal RNAs, and diverse non-coding RNAs) and can potentially have radical effects on the final product. U insertion or C-to-U conversions can lead to the creation of new translation start and stop codons, while U-to-C changes can remove them. A to inosine (I, interpreted as G) conversions in the coding sequence can result in an amino acid change in the resulting protein, and seems to be common in the human brain (see Figure 4.4). A-to-I editing occurs almost exclusively in Alu repeats, primate-specific transposable elements that make up 10 per cent of the human genome. They are common in gene-rich regions and most genes contain multiple copies. Hence the widespread A-to-I editing in humans is likely to be a side effect of the abundance of Alu elements (Barak et al. 2009). Most, but not all, of these repeats occur in introns where they are less likely to interrupt normal gene function. Hence, A-to-I conversions happen mostly in non-coding parts of the transcript where they regulate or interfere with post-transcriptional processing – for example, through the creation or deletion of a splice site, or a change in the secondary structure of RNA with all kinds of downstream effects. Other editing events can affect mRNA transport and stability, among other things, through alterations within introns and 5′ and 3′ untranslated regions (UTRs) (Hundley and Bass 2010). Some known splicing events in exons that code for channel proteins in the brain affect channel properties in neurons. Other consequences of insertion or deletion editing include changing the reading frame of the transcript, so that the sequence is translated as a completely different product.

The extent of editing within coding sequences reaches from singular amino acid substitutions to widespread nucleotide insertions where over 50 per cent of the final mRNA can be the product of editing. Co-transcriptional A-to-I RNA

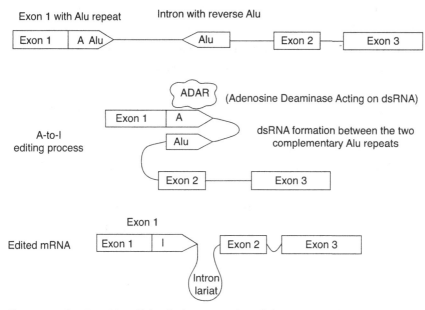

Figure 4.4 A-to-I mRNA editing in humans. The editing enzyme that exchanges an A nucleotide with an I nucleotide works exclusively on double-stranded RNA. Depicted here is an exon–intron double stranded RNA formation, which almost always happens by means of reverse Alu repeats of which most genes contain multiple copies. For simplicity only one editing site is shown, but the number is often much higher (up to twenty editing sites in one dsRNA formation). The mechanism that produces the editing site specificity is not well known. A certain codon bias around the editing site might be involved.

editing, while usually altering highly conserved or invariant coding positions in proteins, may sometimes also correct G-to-A mutations at the DNA level and therefore maintains the original sequence of the gene product. Editing-derived sequence information is often essential for the normal functioning of the organism, as is the case with the A-to-I substitution in the human brain which helps to fine-tune the function of channel proteins.

Editing is a chemical reaction in which enzymes, which are specific to only one kind of editing mechanism, provide the *editing efficiency* by catalysing the deletion, insertion, or substitution reactions. Most substitution editing can be relatively easily catalysed by the known enzymes. The insertion and deletion of nucleotides require a more complicated machinery, especially if it necessitates the reconstruction of the RNA sugar-phosphate backbone (Gott and Emeson 2000; Bass 2001). The chemical details are less interesting here, however, than the more important issue for our thesis of distributed sequence specificity: namely, the second requirement of editing, *site specificity*. The important question is which factors are responsible for guiding the enzyme to the correct nucleotide site in the transcript that will be the target of the editing mechanism. It is likely that both *cis*-acting sequences and *trans*-acting factors will be involved. Editing mechanisms differ considerably, so the answer to this question will require a detailed analysis of each mechanism. In some cases specialised 'guide RNAs' are complementary to the target site. In most cases, however, the exact mechanism behind editing site specificity is not known.

Alternative splicing also requires both *splicing efficiency* provided by some enzyme and *splice site specificity*, provided by the interaction between a myriad of splice site sequences and the many different splicing factors that bind to them.

4.5.3 Translational recoding

Another mechanism which disrupts the colinearity between coding sequences and final product is the *translational recoding* of mRNA. The three different ways through which the translational machinery is able to recode the message are frameshifting, programmed slippage or bypassing, and codon redefinition. In frameshifting, the ribozymes start translating at a different base than the usual start base, either $+1$ or $+2$, so that the whole sequence will be read in a different frame and produce a completely unrelated protein. In slippage or bypassing, the ribosome may slip back to translate a codon again or to insert a new codon if the slip is out of frame, or it may jump forward and not translate some of the sequence. Finally, some organisms have mechanisms which can temporarily assign a new amino acid to a codon or read through a stop codon. There is a database of all known translational recoding cases known as 'RECODE', which lists the DNA sequence, the alternative forms of protein, the particular recoding mechanism involved, and also all the known *trans*-factors and *cis*-sequences involved in its regulation (Baranov et al. 2003).

In this section we have superficially described some very complex and not fully understood molecular processes. We know that this material will have been demanding for many philosophical readers. But it makes an essential conceptual point. *Trans*-splicing, RNA editing, and translational recoding all

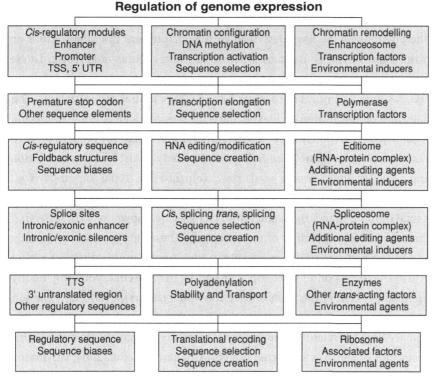

Regulation of genome expression

Cis-regulatory modules Enhancer Promoter TSS, 5' UTR	Chromatin configuration DNA methylation Transcription activation Sequence selection	Chromatin remodelling Enhanceosome Transcription factors Environmental inducers
Premature stop codon Other sequence elements	Transcription elongation Sequence selection	Polymerase Transcription factors
Cis-regulatory sequence Foldback structures Sequence biases	RNA editing/modification Sequence creation	Editiome (RNA-protein complex) Additional editing agents Environmental inducers
Splice sites Intronic/exonic enhancer Intronic/exonic silencers	*Cis*, splicing *trans*, splicing Sequence selection Sequence creation	Spliceosome (RNA-protein complex) Additional editing agents Environmental inducers
TTS 3' untranslated region Other regulatory sequences	Polyadenylation Stability and Transport	Enzymes Other *trans*-acting factors Environmental agents
Regulatory sequence Sequence biases	Translational recoding Sequence selection Sequence creation	Ribosome Associated factors Environmental agents

Figure 4.5 Factors with sequence specificity in eukaryotic genome expression. The middle column shows the stages of transcriptional and translational processing from the packaged DNA through RNA to the amino acid sequence. The left column lists different *cis*-regulatory sequences involved at the different stages of genome expression, while the right column shows *trans*-acting factors at each stage.

show that *the sequence of gene products, even if defined narrowly as the linear order of their elements, derives from regulatory mechanisms as well as the DNA sequence from which the product is initially transcribed.*

The moral of this section and 4.4 is that the relationship between DNA and gene product is indirect, mediated, and subject to regulated contributions by other sequence-specifying agents (see Figure 4.5). Crick information is not restricted to the context in which Crick originally identified it: namely, the sequence from which mRNAs are transcribed. One obvious response to this is to admit that additional Crick information is derived from these mechanisms, but insist that this information ultimately comes from DNA sequences elsewhere in the genome. We deal with this objection in Chapter 5.

4.6 Molecular epigenesis

The phenomena we have described in this chapter are not singularities or rarities but business as usual, at least in multi-cellular eukaryotes. Latest estimates place the percentage of alternatively spliced human genes close to 100. A significant number of genes have 5,000 potential splice variants. The flagship example of alternative splicing is the *Drosophila* cell adhesion molecule (*Dscam*) gene with potentially 38,016 temporally and spatially regulated splice variants, of which more than half are confirmed to exist in nature (Celottoa and Graveley 2001; Rowen et al. 2002; Leipzig et al. 2004; Kapranov et al. 2005; Graveley 2005). Editing is also ubiquitous. In some plant mitochondria a total of more than a thousand C-to-U changes are known to occur, affecting the entire RNA population of mitochondria. These substitutions are mostly within the first two positions of codons, hence producing a different amino acid in the corresponding protein. RNA editing in some eukaryotes can modify up to 50 per cent of the adenosine residues in a transcript (Gott and Emeson 2000). Some forms of editing are critical for normal brain function in humans and are very prevalent in human cells, particularly some brain tissue, with thousands of genes as suspected targets (Athanasiadis et al. 2004; Paz-Yaacov et al. 2010; Park et al. 2012). A recent study of the architecture of the human transcriptome paints 'a picture of a highly overlapping, complex, and dynamic nature of the human transcriptome, where one base pair can be part of many transcripts emanating from both strands of the genome. The data further suggest that base pairs normally thought to contribute to transcripts from different genes can be joined together in a single RNA molecule' (Kapranov et al. 2005; see also Kampa et al. 2004; Cheng et al. 2005).

What do these findings tell us about the nature of genes and genomes? We suggest that they support three interrelated theses: genetic underdetermination and amplification, distributed causal specificity by means of regulated recruitment and combinatorial control, and molecular epigenesis.

4.6.1 Genetic underdetermination and amplification

We noted at the beginning of the chapter that the human proteome (the number of proteins found in human cells) outnumbers the protein-coding genes in the human genome by at least one order of magnitude. Findings of this kind suggest that gene products are underdetermined by the coding sequences from which their precursor molecules are transcribed; transcriptional and post-transcriptional processes 'amplify' the coding potential of the coding regions themselves:

> There is increasing awareness that multiple, often overlapping mechanisms exist for amplifying the repertoire of protein products specified through the mammalian genome. An expanding array of processing and targeting mechanisms is now emerging, each representing a potentially important restriction point in the regulation of eukaryotic gene expression, and each expanding the possibilities specified by the literal code of the genome. (Davidson 2002, 291)

To say that these factors 'amplify' the coding potential of the genome is to say that they partly determine which products the genome codes for. The specificity for the sequence of a final gene product is distributed between the coding DNA and these other factors: *cis*-acting sequences, *trans*-acting regulators, environmental signals, and the contingent history of the cell (the 'cellular code') (see Figure 4.5 for an overview). *These are alternative sources of sequence information.*

4.6.2 Distributed causal specificity by means of regulated recruitment and combinatorial control

Because all the factors just listed are causally specific difference makers with respect to the sequence of gene products, that specificity is *distributed* rather than localised in the coding regions of the genome as Crick's sequence hypothesis suggested. The ability of this array of factors to amplify the coding potential of the genome in turn depends on two key factors: *regulated recruitment* and *combinatorial interaction* (Ptashne and Gann 2002; Buchler et al. 2003). In eukaryotes, and to some extent in prokaryotes, the regulation of gene expression works by means of the regulated recruitment of *trans*-acting factors (proteins, RNAs, and metabolites) into larger complexes and their recruitment in turn to *cis*-acting sequence modules, so that the specificity of an enzyme, a sequence, a transcription or splicing factor comes to depend on its combinatorial association with others. The modular organisation of genes and of *cis*-regulatory sequences and the modular organisation of *trans*-acting factors into functionally distinct subunits (DNA binding sites, protein-protein and protein-RNA recognition sites, and catalytically active sites) contributes to the capacity for combinatorial control. This combinatorial complexity, which massively expands the repertoire of coding regions, resolves the 'N-value' paradox that the number of protein-coding genes of an organism doesn't seem to be correlated with its complexity (Claverie 2001; Harrison et al. 2002).

New metaphors have been suggested to capture this emerging picture of genome expression: 'the sociable gene' conveys the interactivity, fluidity, and dynamics of the genomic system; Lenny Moss talks of 'ad hoc committees' of molecules convened on the basis of the history of the cell and its interaction with the environment to describe the control of genome transcription; the 'cooperative genome' invokes the continual cooperation between multiple factors involved in most cellular processes from protein synthesis to development (Moss 2003; Turney 2005; Weiss and Buchanan 2009). We will describe the situation in a less metaphorical way as 'molecular epigenesis' (Stotz 2006a, 2006b; a term first used by Burian 2004).

4.6.3 Molecular epigenesis

In the 1950s Conrad Waddington drew an analogy between different views of the role of genes and the ancient rival theories of preformation and epigenesis:

> Some centuries ago, biologists held what are called 'preformationist' theories of development. They believed that all the characters of the adult were present in the newly fertilized egg but packed into such a small space that they could not be properly distinguished with the instruments then available. If we merely consider each gene as a determinant for some definite character in the adult (as when we speak loosely of the 'gene for blue eyes, or for fair hair'), then the modern theory may appear to be merely a new-fangled version of the old idea. But in the meantime, the embryologists [. . .] have reached a quite different picture [. . .] This is the theory known as epigenesis, which claims that the characters of the adult do not exist already in the newly fertilized germ, but on the contrary arise gradually through a series of causal interactions between the comparatively simple elements of which the egg is initially composed. There can be no doubt nowadays that this epigenetic point of view is correct. (Waddington 1952, 156)

It is sometimes said that molecular biology is a partial vindication of preformationism. Although the structure of body parts is not preformed, the structure of the molecular parts, proteins, is preformed in the DNA (Gould 1977;

Godfrey-Smith 2000a; Godfrey-Smith 2001). Our arguments in this chapter suggest that molecular preformationism, like its morphological predecessor, is mistaken. Like morphological structures, biomolecules are constructed by an epigenetic process. Multiple factors, none of which contain a full representation of the molecule, are brought together in processes regulated by the larger system of which they are part. The Crick information manifest in a biomolecule is produced by an 'ontogeny of information' (Oyama 1985).

An obvious response is that while any particular coding sequence may only partially specify its product, the genomic sequence as a whole, including the cis-acting regulatory sequences and the genome regions from which the transacting factors are themselves transcribed, fully specifies that product. Hence the Crick information for gene products remains entirely within the genome, albeit globally and not locally. However, complex eukaryotes have no default transcriptional activation or splicing pattern. Specific difference-making RNA and protein factors need to be recruited or activated by external inducers. These molecules undergo crucial changes in shape in response to these signals, which render them active and impose their causal specificity (Ptashne and Gann 2002, 6–7). One can indeed say that environmental signals are the 'drivers' of gene expression (Istrail and Davidson 2005). The significance of this is that these molecules relay specific difference-making (instructive) environmental information to the genome. While it is common to regard the role of the extra-cellular environment in gene expression as merely permissive, in many cases signals from the environment provide specificity for gene products. Through mechanisms like those described above, they co-specify the linear sequence of the gene product together with the target DNA sequence. Developmental biologist Scott Gilbert makes this point in the paper whose title we borrowed for this chapter, 'The Reactive Genome':

> Organisms have evolved [a reactive genome] to let environmental factors play major roles in phenotype determination [...] In instructive interactions, a signal from the inducer initiates new patterns of gene expression in responding cells [...] It is usually assumed that the developing organism's environment constitutes a necessary permissive set of factors, whereas the genome provides the specificity of the interaction. In phenotypic plasticity, however, the genome is permissive and the environment is instructive. (Gilbert 2003, 92)

Gilbert restricts this claim to the development of phenotypically plastic traits of the organism, traits that take different forms depending on the environment. We believe that it holds much more widely, as we will show in 5.6 with our concept of the 'developmental niche' for genome expression.

4.7 Systems biology: from reduction to integration

> Why then (in an organism where we know so much about its biochemistry, physiology and cell biology) should it be a problem to identify the biological subsystems that must be fully characterized and built into a comprehensive model of the eukaryotic cell? This problem arises because we have previously studied these biological systems in isolation and in a rigorously reductive fashion. Now, we must study them as parts of an integrated whole. (Oliver 2006, 478)

The term 'systems biology' is used to describe a variety of research intended to respond to the challenges of biology in the postgenomic era: namely, vast bodies of data and explanatory and pragmatic challenges which require coming to terms with the results of many molecular mechanisms operating simultaneously and interactively in a living system. In response, systems biology attempts to understand living systems through the study and rigorous simulation of system-level organisation and dynamic interaction. However, this description glosses over some fundamental differences in approach. Maureen O'Malley and John Dupre (2005) have drawn a useful distinction between 'pragmatic systems biology', a more practical perspective on modelling large-scale interactions, and 'systems-theoretic biology', a more theoretical perspective which seeks principles that apply to biological systems in general. These two perspectives roughly align with another distinction, that between 'bottom-up' and 'top-down' approaches in systems biology. Bottom-up approaches assemble a description of the system by characterising its components and documenting their interactions. Top-down approaches use models to identify principles of system organisation which can then be used to explore real systems. Dennis Noble, one of the founders of systems biology, remarks:

> The consensus is that it should be 'middle-out', meaning that we start modeling at the level(s) at which there are rich biological data and then reach up and down to other levels. (Noble 2002, 1980)

Some systems biologists retain a connection to the general systems theory of Ludwig von Bertalanffy, who also coined the term 'systems biology' (Wolkenhauer 2001). Bertalanffy hoped to integrate the life sciences by discovering universal principles of systems organisation, and some still have this aim and believe the time is ripe for it. 'The delay between the early pronouncement of the theory and the work presently assembled was necessary, primarily to accumulate sufficient descriptions of the parts to enable a reasonable reassembly of the whole' (Chong and Ray 2002, 1661).

Systems biology has been characterised by slogans such as 'from sequence to biology', 'to the centre of biology', or 'from genomes to systems', all of which suggest a move from more reductionist to more integrative approaches (Lander and Weinberg 2000; Stein 2001; Oliver 2006). At the very least, the focus of systems biology is not on characterising individual system components in greater detail, but on understanding the integrated operation of networks of genes, protein, gene regulatory mechanisms, metabolic products, and their interactions.

> Twentieth century biology triumphed because of its focus on intensive analysis of the individual components of complex biological systems. The 21st century discipline will focus increasingly on the study of entire biological systems, by attempting to understand how component parts collaborate to create a whole. For the first time in a century, reductionists have yielded ground to those trying to gain a holistic view of cells and tissues. (Lander and Weinberg 2000, 1781)

But systems biology is not just a move of the pendulum back to holism; it actually embraces and builds on reductionist research strategies and their successes. In this respect it fits the description of mechanistic anti-reductionism that we gave in 3.5. It may therefore be better to describe it as 'integrationist' rather than as holistic:

> An integrationist, using rigorous system-level analysis, does not need or wish to deny the power of successful reduction. Indeed he uses that power as part of his successful integration [...]

> Integrative systems biology is just as rigorous and quantitative as reductionist molecular biology. The only difference is that it accepts that causality goes from higher to lower levels as well as upwards. (Noble 2006, 66, 77)

It is noteworthy that Noble identifies top-down causation as a distinctive feature of explanation in systems biology. The leading advocate of reductionism in contemporary philosophy of biology, Alexander Rosenberg, also sees this as a key issue separating reductionists and anti-reductionists (Rosenberg 2006). Rosenberg appeals to the famous 'causal exclusion' argument (Kim 1993). If living systems are exhaustively composed of physical constituents, then the evolution of the system in accordance with the laws of physics will be sufficient to produce all the changes that occur in the system over time. There is no room for any more causation. Not only can higher-level structures like cells and tissues not cause any changes in lower-level structures, they cannot even cause changes at their own level. Not only can mechanical stress on a bone not cause a change in gene expression patterns in osteoblasts, it cannot cause the bone to break! The trouble with the causal exclusion argument in the present context is that it uses an ontological conception of causation to attack epistemic anti-reductionism. It treats causation as something ontic which moves the world on from one stage to the next, and concludes that there is no need – and therefore no place – for any more causation than is captured in fundamental physics. But, as we noted in Chapter 3, epistemic anti-reductionism is not an ontological claim. It is a family of claims about the relationship between different scientific domains and their bodies of knowledge. In this context, a more appropriate conception of causation is something like that offered by Woodward (2003). Causation is a certain kind of a relationship between variables: namely, a relationship that can be used to make things happen in the system. There is no reason to think that the causal exclusion principle applies to such a conception of causation.

In fact, what anti-reductionists mean by top-down causation is something quite mundane. Craver and Bechtel (2007) identify top-down causation with the constraints placed on the behaviour of parts by their interactions with the other parts of the system that contains them. The constituents of a system behave differently from how they would behave in isolation or as parts of another system. This is explained by the organisation of the system. This is a non-reductionist explanation because it does not result (solely) from identifying the parts of the system and characterising those parts, but from analysing how the parts are organised and the phenomena that arise from their being so organised. The most controversial version of top-down causation, and it is surely not very controversial, is when the activity of the part is explained by what Bechtel and Abrahamsen term a dynamic mechanistic explanation.

When it is necessary to use a dynamical system model to explain how the mechanism changes over time it may be that the only explanation that can be given for the state of a component part at a time is given by exhibiting the dynamics of the system as a whole (Bechtel and Abrahamsen in press).

As Brigandt and Love (2008) have pointed out, anti-reductionists often make top-down causation seem more mysterious than it is by focusing on the constitutive causal relations between a higher-level entity and its lower-level parts, rather than dynamic causal processes that bridge higher and lower levels. The parts cause the whole to be the way it is, so how can the whole simultaneously cause the parts to be the way they are? First of all, the part–whole relationship is instantaneous: the whole determines the part as much as the other way around. But top-down causal explanations do not have to be synchronous in this way – in fact they generally are not. When we say that mechanical stress on the bone causes gene expression in the osteoblasts, we do not mean gene expression just as I stress the bone, but gene expression immediately afterwards. This is 'top-down' causation because the relationship between a macro-level torsional force (for example) and activity at some locus on the genome of an osteoblast only exists because of how very many constituents of the system are arranged with respect to one another.

When we look at the kinds of claims about top-down causation made by anti-reductionists in biology it is clear that they can be interpreted in these mundane ways. For example, in a discussion of programmed cell-death, Morange (2008, 105) remarks that it has been accepted for a long time 'that individual cells surrender part of their identity to the organism as a whole'. 'The evolution of multicellular organisms can now be seen [. . .] to have introduced a top-down system of control in which the fate of the organism dictates the fate of individual cells' (107). There is no need to postulate anything more mysterious than (perhaps dynamic) mechanistic explanations which depend on features of a system's organisation to make sense of remarks like this.

Noble's claim that understanding living systems requires coming to terms with top-down causation chimes with the themes that have emerged in this chapter. It fits the ideas about genome function that we have sketched. The focus of postgenomic biology is no longer on a single sequence and its exclusive causal specificity but on the network of regulatory mechanisms of genome expression with distributed specificity. It also fits the new ideas about genome structure that we have introduced. The postgenomic gene is an essentially top-down conception. A set of sequences is a gene because of the way in which it is used by the cell, not because of its intrinsic structure. Our own favourite example is *nad1*, a plant mitochondrial gene which is *trans*-spliced together from five different genomic locations and then requires heavy mRNA editing to produce a functional transcript (Chapdelaine and Bonen 1991; Stotz et al. 2006).

A theme in many discussions of systems biology is that new principles of system organisation and functioning will be needed in order to explain the activity of living systems. Cataloguing the parts of the system and the specific interactions in which each part participates may not be enough. This can be understood in terms of another classic anti-reductionist theme, which is the context-dependence of the functioning of molecular features, which lead to a one-to-many relationship between a molecular feature and its effects in different contexts. One and the same transcription factor may act as an activator or inhibitor depending on which regulatory sequence it binds or which other factors it interacts with. There is a profound difference between the *molecular function* of a molecule, which it owes to its molecular structure, and the realised *cellular function* which it owes to its context and its interaction with other entities (Helden et al. 2000). This distinction poses something of a challenge to the very idea of specificity. It suggests that specificity may not be an intrinsic property of a sequence or a structure but a contextual property, as in the many examples of regulated recruitment and combinatorial control above. This has profound effects for the translation of biological and biomedical research into applications:

> [E]ven when we understand function at the protein level, successful
> intervention, for example, in drug therapy, depends on knowing how a
> protein behaves in context, as it interacts with the rest of the relevant cellular
> machinery to generate function at a higher level. Without this integrative
> knowledge, we may not even know in which disease states a receptor, enzyme,
> or transporter is relevant. (Noble 2002, 1678)

The rise of systems biology strongly supports the picture we sketched in 3.5. Contemporary molecular biological research is both reductionist and integrative. It retains a commitment to the reductive strategy of decomposing systems into their components and characterising those components, but its explanatory strategy relies on the functional organisation of systems and the top-down effects of that organisation. The new identity that the gene acquires in this research context, the postgenomic gene, is an entity defined top-down by the system of which it is a part. The genome is no longer an inner controller, but has become a reactive structure embedded in a wider environment. It is this system around the reactive genome that systems biology aims to understand. In the next chapter we will turn our attention to the components of that system that lie outside the genome.

Further reading

A good introduction to the regulation of transcription in both prokaryotes and eukaryotes, and from which we derived the phrases 'regulated recruitment' and 'combinatorial control', is *Genes and Signals* (Ptashne and Gann 2002). An excellent popular introduction to the systems-biological understanding of genes and their interaction within an organism is *The Music of Life: Biology beyond Genes* (Noble 2006). In *Embryology, Epigenesis and Evolution* Jason Scott Robert (2004) applies an epigenetic perspective to today's understanding of development. The classic account of how to deal with complexity at the methodological level is *Discovering Complexity* (Bechtel and Richardson 1993).

5   Outside the genome

5.1 Introduction

> As the past 70 years made abundantly clear, genes do not control
> development. Genes themselves are controlled in many ways, some by
> modifications of DNA sequences, some through regulation by the
> products of other genes and/or by [the intra- or extra-cellular] context,
> and others by external and/or environmental factors.
>
> (Hall 2011, 9)

In the previous chapter we saw that regulated expression of the genome
depends on many other factors besides the DNA sequence. We established
that these other factors are not merely background or 'permissive' causes
which allow the expression of the Crick information contained in a coding region. Instead, regulatory mechanisms differentially activate and select
coding sequences depending on the context, and thus contribute additional
Crick information, and some mechanisms create Crick information for which
there is no underlying sequence. Crick information, we argued, is distributed
between the coding regions in the genome and the regulatory mechanisms
that use those coding regions to make products. Our focus in this chapter is
on the role of the environment, the 'developmental niche', in that regulatory
machinery.

Many biologists familiar with the mechanisms described in Chapter 4 think
that their action can be traced back to other regions in the genome, so that
the genome as a whole remains the sole source of biological specificity. It
is certainly true that regulatory gene products and the genome regions to
which they bind are at the heart of most of the mechanisms we described.
But this is nowhere near sufficient to establish that the genome is the sole
source of specificity. That idea is more difficult to state precisely, let alone

to defend, than most of its advocates realise. We think that most versions of the idea equivocate between two or more of the senses of 'genetic information' that we will explore in Chapter 6. Sticking strictly to the idea of Crick information – the determination of the order of elements in a gene product – clarifies the situation. Most of the regulatory molecules we described contain Crick information from genome sequences because they were transcribed from those sequences. But they do not contribute to specifying gene products by unloading this cargo of information. For example, when a regulatory molecule changes the conformation of a transcription complex by binding to it, this does not involve transferring the Crick information from the sequence of the regulatory molecule to the sequence of the molecule that is being transcribed. The sequence of the regulatory molecule and the molecule it regulates do not correspond to one another in the right way. We have argued that such mechanisms contribute Crick information to the product, but this does not come from the coding region from which the regulatory molecules in the mechanism were transcribed. We do not see that the information has to 'come from' anywhere, except perhaps from evolutionary design, as we will discuss in Chapter 6. By acting as a causally specific difference maker with respect to the sequence of the product, a mechanism provides Crick information for that product which does not come from the coding sequence being regulated. It would require substantial theoretical work to show that this Crick information is encoded by the linear specificity of other genome sequences.

In this chapter we will provide reasons to doubt that distributed specificity can all ultimately be traced back to the genome. We do this by showing how the environment, acting through the regulatory mechanisms described in Chapter 4, plays an instructive role in regulating gene expression. We examine a number of existing approaches to this issue. In the next two sections we look at the new field of epigenetics. We discuss the complex and contested meanings of the term 'epigenetics', explain its molecular underpinnings, including DNA methylation, histone modifications, cytoplasmic inheritance, and non-coding RNAs (ncRNAs), and examine some of the practical applications of epigenetics in medical research. In 5.4 we look at research into 'parental effects', an alternative, very different, way of looking at many of the very same phenomena that are studied in epigenetics. In 5.5 we examine other research agendas that focus on overlapping subsets of these phenomena: developmental plasticity; ecological developmental biology; and epigenetic or, as we prefer to call it, 'exogenetic' inheritance. Finally, in 5.6 we introduce the idea of a 'developmental niche' and suggest that it is a good framework for integrating ideas and results from all these research agendas. The developmental niche is the set of environmental and social legacies that make possible the regulated expression of the genome during the life cycle of the organism (West and King 1987).

As well as continuing the argument of Chapter 4 about distributed specificity, the material presented in this chapter is of interest for two other reasons. First, it has implications for evolutionary theory, which are the topic of Chapter 8. Because we devote a whole chapter to these implications later in the book we will mostly leave them implicit at this stage. Second, this material undermines the common oppositions between nature and nurture, innate and acquired, biology and environment. We will return to this idea in the conclusion of the chapter.

5.2 Epigenetics

> The term 'epigenetics' is derived from the process of epigenesis. As a continuation of the concept that development unfolds and is not preformed (or ordained), epigenetics is the latest expression of epigenesis. (Hall 2011, 12)

The idea of epigenesis is sometimes traced back to Aristotle's *On Generation*, and sometimes to William Harvey's 1650 book of the same title. The long-running dispute between epigenesis and the preformation theory of development broke out at the end of the seventeenth century. Harvey had convinced scientists that all animals develop from an egg. Some scientists also accepted that the 'seminal animalcules' (spermatozoa) discovered by Antony van Leeuwenhoek in the 1670s merged with the egg, or at least stimulated the egg to begin developing. The debate was over how the egg gives rise to the animal. Epigenesis states that the contents of the egg are relatively simple and that the operation of natural laws on these simple ingredients leads to increased complexity. The term 'epigenesis' derives from the Greek for 'new origin'. Preformationism states that the structure of the organism is preformed in the egg (or in the sperm, for which the egg is a 'nest'). The egg is as complex and highly ordered as the adult that arises from it. Epigenesis and preformation were debated by students of embryology throughout the eighteenth century (Maienschein 2005).

Preformationism is often mocked as the view that the egg contains a little man curled up ready to grow. But in the hands of leading eighteenth-century exponents like the Swiss naturalist Charles Bonnet it was no more absurd than the view that development is the expression of a genetic programme. According to Ernst Mayr (1961), natural selection has written a genetic programme which constructs the organism (see 6.4). Bonnet held that God has designed an exquisitely complex Newtonian mechanism that unpacks itself into an organism. Both are attempts to explain what guides development. Conversely, epigenesists are often said to have believed in a vital force or life force that could turn a soup of particles into an organism. But enlightenment epigenesists like Denis Diderot and Pierre de Maupertuis were the hard-line materialists of their day. Living systems, they insisted, are nothing more than complex collections of particles governed by the laws of physics and chemistry. Since the laws then known were manifestly insufficient to do the job, they postulated that new laws would be discovered as science progressed. This was hardly an unreasonable view when chemistry had only just started to take form and electricity was a new and barely understood phenomenon. But nothing happened in the eighteenth century to change the view of one early preformationist that 'all the Laws of Motion which are as yet discovered, can give but a very lame account of the forming of a Plant or Animal.' (Garden 1691, 476).

In the nineteenth century, preformationism was recast as the more general doctrine of predeterminism, the idea that development consists of an orderly progression of qualitative change to a predetermined endpoint. What predeterminism has in common with the old preformationism is the view that the environment of the egg and the physical laws are non-specific or permissive factors while all the specific or instructive factors are inside the egg, or the nucleus, or the genome (Gottlieb 2001). Mayr's view that development is guided by a genetic programme is a predeterminist view (Mayr 1961).

Biologists unhappy with the predeterminist flavour of the conventional view of the role of genes in development have often compared themselves to the old epigenesists, as we will see in a moment. They understand development as a process of qualitative change in which there is an orderly emergence of novel traits during development without recourse to a pre-existing plan, and recognise a much greater instructive role for the environment in development (Gottlieb 1992; Michel and Moore 1995; Oyama et al. 2001; Robert 2004).

Box 5.1 Definitions of epigenetic

Epigenesis: the idea that the outcomes of development are created in the process of development, not preformed in the inputs to development; 'epigenetic' can be used in these senses:

Epigenetics (broad sense – Waddington 1940): the study of the causal mechanisms by which genotypes give rise to phenotypes; the integration of the effects of individual genes in development to produce the 'epigenotype'.

Epigenetics (narrow sense – Nanney 1958): the study of the mechanisms that determine which genome sequences will be expressed in the cell; the control of cell differentiation and of mitotically and sometimes meiotically heritable cell identity.

Epigenetic inheritance (narrow sense): the inheritance of genome expression patterns across generations (e.g. through meiosis) in the absence of a continuing stimulus.

Epigenetic inheritance (broad sense): the inheritance of phenotypic features via causal pathways other than the inheritance of nuclear DNA. We refer to this as 'exogenetic inheritance' (West and King 1987).

5.2.1 From epigenesis to epigenetics

The term epigenetics was introduced by Waddington (1940), through the fusion of 'epigenesis' and 'genetics', to refer to the study of the processes by which the genotype gives rise to a phenotype. Waddington emphasised that epigenetics is a search for causal mechanisms, and suggested that existing knowledge from experimental embryology supported a view of how genes are connected to phenotypes broadly in line with the older idea of epigenesis. The term is still used in this broad sense today, but has also acquired the related but much narrower sense of the 'study of changes of gene expression, which occur in organisms with differentiated cells, and the mitotic [and/or meiotic] inheritance of given patterns of gene expression' (Holliday 1994, 453).

So today the term 'epigenetics' has two distinct meanings (Haig 2004; see Box 5.1). The broader sense goes back to Waddington and has always appealed to developmental biologists, for some of whom 'epigenetics refers to the entire series of interactions among cells and cell products which leads to morphogenesis and differentiation' (Herring 1993, 472; cited in Haig 2004, 1; cf. Holliday

2006). The evolutionary developmental biologists Benedikt Hallgrimsson and Brian Hall are using Waddington's original sense of epigenesis when they claim that epigenetics links the genotype to the phenotype in *both* development and evolution:

> Epigenetics is the study of emergent properties in the origin of the phenotype in development and in modification of phenotypes in evolution. Features, characters, or developmental mechanisms/processes are epigenetic if they can only be understood in terms of interactions that arise above the level of the gene as a sequence of DNA. (Hallgrimsson and Hall 2011, 1)

To clarify the relationship with the narrower, molecular sense of epigenetics, they say that:

> Methylation and imprinting of gene sequences are examples of epigenetics at the level of the structure and function of the gene. (1)

However, many molecular biologists now understand epigenetics primarily in the narrow sense, as the study of changes in gene expression that are mitotically heritable (via somatic cells) or meiotically heritable (via germ cells), and that do not entail changes in DNA sequence. According to Haig, this sense originates from David Nanney's use of the term 'epigenetic control systems' regulating the expression of genetic potentialities, which he hypothesised to be auxiliary to genetic control systems and that operate through a steady-state system of self-regulating metabolic patterns (Nanney 1958; Haig 2004). Writing in the same year that Crick stated the sequence hypothesis, Nanney discussed the idea that specificity is transmitted in genetic templates and argued that this necessitated regulatory mechanisms to determine which templates were used at a particular point in the life of the cell:

> This view of the nature of the genetic material, while certainly not established in detail, finds much support in experimental studies and gains great strength from its simplicity. It permits, moreover, a clearer conceptual distinction than has previously been possible between two types of cellular control systems. On the one hand, the maintenance of a 'library of specificities,' both expressed and unexpressed, is accomplished by a template replicating mechanism. On the other hand, auxiliary mechanisms with different principles of operation are involved in determining which specificities are to be expressed in any

particular cell [...] To simplify the discussion of these two types of systems, they will be referred to as 'genetic systems' and 'epigenetic systems.' The term 'epigenetic' is chosen to emphasize the reliance of these systems on the genetic systems and to underscore their significance in developmental processes. (Nanney 1958, 712)

Nanney cited Waddington as the source of the term 'epigenetic' but his idea of an epigenetic system is much more specific than Waddington's and it was widely adopted by researchers interested in mechanisms of cell-line heredity. Haig suggested that the paper 'The Inheritance of Epigenetic Defects' (Holliday 1987) led to much wider use of the term 'epigenetic' in the sense of the heritable control in gene expression, and the author, in his own history of the term, seemed to agree (Haig 2004; Holliday 2006).

5.3 Epigenetic mechanisms

In a multi-cellular organism every cell has the same DNA but there are many cell types, each with a different pattern of gene expression. The modern study of epigenetics at the molecular level has been primarily concerned with identifying the mechanisms that allow cells to remember their cell identity, and to pass on that identity when they divide.

> As long as a transcriptional response is self-sustaining in the absence of the originating stimulus, it can be categorized as epigenetic. This can be achieved by self-propagating, *trans*-acting mechanisms or by *cis*-acting molecular signatures physically associated with the DNA sequence that they regulate. (Bonasio et al. 2010, 612)

The *trans*-acting mechanisms include a range of cellular processes, the most common of which are the maintenance of transcriptional states through steady states: that is, feedback loops of networks of transcription factors that regulate their own or each other's genes. Since all organisms use this mechanism this may be the oldest epigenetic mechanism. Another *trans*-acting mechanism employs regulatory non-coding RNAs which regulate gene expression, either by themselves or in combination with DNA or histone modifications. Both transcription factors and ncRNAs are transmitted by the partitioning of the cytoplasm in cell division and resume their functions in the daughter cells. Some maternal and paternal gene products are also inherited to the cytoplasm and regulate gene expression during the first few cell divisions of the embryo.

These mechanisms cannot be redeployed from one gene network to another. Each network that needs to be stabilised has to develop its own feedback system. But many organisms have invented ways to encode their epigenetic state in *cis*-acting epigenetic signals. In contrast to *trans*-acting mechanisms, these are physically associated with the genetic material (the chromosomes) with which they are inherited. At least some *cis*-acting signals need to be established only once and are then easily maintained by a few mechanisms that work for all gene networks. The most important of these general-purpose signals are covalent modifications of the DNA itself, mainly the methylation of the nucleic acid base cytosine at CpG sites to create 5-methylcytosine (see Figure 5.1). The attachment of a methyl group to one or more C residues in promoter sequences decreases or fully prohibits the binding of transcription factors to the promoter where it would recruit the enzyme RNA polymerase. Transcription is either down-regulated or cannot take place at all, and the gene is silenced.

Methylated DNA binds with methyl-CpG binding domain (MBD) proteins, which are either associated with, or recruit, large protein complexes with chromatin-modifying abilities. Chromatin consists of the DNA and associated histone proteins (see bottom of Figure 5.1). Around 150 base pairs of DNA are wrapped around a core histone octamer made up of two copies each of the histone molecules H2A, H2B, H3, and H4. This complex forms the building block of chromatin and is called a nucleosome. The state of the chromatin defines how loosely or strongly the DNA is attached to its histone core, which renders the attached DNA either accessible or inaccessible to the transcriptional machinery. Different chromatin states are associated with a range of diverse post-translational modifications of the histone proteins at distinct amino acid residues on their amino (N)-terminal tails which face outward from the nucleosome (see Figure 5.1, middle section). We don't know the function of all of these modifications but hyperacetylation is usually thought to open up the chromatin in order to increase transcription, and hypoacetylation promotes condensation and decreased gene activity. Other modifications such as methylation or phosphorylation can correlate with either activation or repression, depending on which locations in which histones are modified and in which combination. The sum of specific combinations of histone modifications at a promoter has been called a 'histone code' (Jenuwein and Allis 2001).

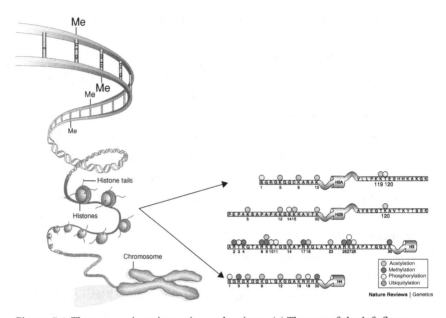

Figure 5.1 The two main epigenetic mechanisms. (a) The top of the left figure shows the unwound double DNA strand with *DNA methylation*. A methyl group (Me) is added to certain DNA bases, normally C nucleotides in the promoter region, to repress gene activity. The middle part of the figure shows the DNA wrapped around histone proteins that form the nucleosomes. In *histone modification* a combination of different chemical groups can attach to certain amino acids that form the tails of the histones. These modifications influence the way the DNA is attached to the nucleosome, which affect the activity of the DNA. (b) The histone code. The figure on the right shows the modification in more detail. Amino acid residues of histone molecules – especially those located at their amino (N)-terminal tails – are subject to various post-translational modifications, including methylation, acetylation, phosphorylation, ubiquitylation, sumoylation, citrullination, and ADP ribosylation. The combination of the different kinds of modifications at the different amino acids in the histones are sometimes described as the 'histone code'. (Reprinted by permission from Nature Publishing Group. 5.1a: J. Qiu, 2006. Epigenetics: unfinished symphony. *Nature* 441: 143–5. 5.1b: Mikhail Spivakov and Amanda G. Fisher, 2007. Epigenetic signatures of stem-cell identity. *Nature Reviews Genetics* 8: 263–71.)

The mechanisms by which DNA methylation gets transmitted during cell division are well understood. DNA replication is 'semi-conservative', meaning that the two single strands of the original DNA each become half of a newly created double strand of DNA. If the original strands were methylated, the new double strands are semi-methylated. DNA methyltransferases are enzymes responsible for attaching methyl group to cytosine residues. So-called main-tenance methyltransferase recognises semi-methylated cytosines and restores them to fully methylated status by attaching a methyl group to the cytosine next to a guanine that is paired with a methylated cytosine. Only cytosine with a guanine as neighbour gets methylated. The mechanisms by which his-tone modifications are or could be inherited are less clear. Some argue that there is no evidence for the inheritance of chromatin states from one cell to the other (Ptashne 2007). Others have proposed possible mechanisms of the propagation for at least some histone-associated signals: for instance, through the transmission of intermediary epigenetic signals such as RNAs (Bonasio et al. 2010). The faithful transmission of X chromosome inactivation during mitosis seems to suggest the existence of such mechanisms. In female mam-mals one copy of the X chromosome is randomly inactivated in each cell by compressing the chromatin into an inactive form. This copy stays inactivated in all the descendants of that cell.

A topic of intense interest at present is the signals that promote DNA methylation and histone modifications in the first place. Indirect evidence strongly suggests that methylation states change in response to environmental signals. One possible scenario is the activation of hormones through sensory stimuli that either function as transcription factors or activate transcription factors, which in turn are able to recruit chromatin-remodelling proteins or the enzyme that methylates C residues (see, e.g., Meaney and Szyf 2005).

5.3.1 Genomic imprinting

Diploid organisms inherit one chromosome from each parent. With the excep-tion of the sex chromosomes, nothing about the DNA indicates whether it comes from a male or a female parent. However, in mammals, insects, and plants, certain alleles are epigenetically marked (imprinted) according to the sex of the parent, so that only the allele inherited from one parent is active, while the other allele is permanently silenced. In humans we have about 200 imprinted genes, most of which regulate early development. Most imprint-ing involves DNA methylation, but histone modifications and ncRNAs may also be involved in some organisms. Multiple hypotheses have been proposed for the evolution of imprinting, most notably the *parental conflict hypothesis* (Haig 2000). Briefly, alleles derived from the mother are guaranteed to be closely related to one another, but alleles derived from fathers have no such guarantee, because there may be more than one father. So alleles derived from the mother should behave in a way that allows the mother to allocate resources to her other offspring, whereas alleles derived from the father should attempt to monopolise resources. In line with this hypothesis several genes that promote growth are imprinted by the mother, while other genes that inhibit growth are imprinted by the father. However, not all imprinted genes are involved in growth and development, and many different genes are imprinted in different organisms, hence one hypothesis is unlikely to explain all instances (Reik and Walter 2001a; Reik and Walter 2001b; McEachern and Loyd 2011).

In mammals many imprinted genes occur in clusters that share regulatory regions. Imprinted regions contain not just protein-coding but also non-coding RNA and regulatory antisense genes. The best-studied imprinted genes in humans are *H19*, a ncRNA gene where the paternal allele is imprinted, and the *Insulin-like growth factor 2 (Igf2)*, a protein-coding gene where the maternal allele is imprinted. A single shared control region regulates these reciprocally imprinted genes. When non-methylated, as in the chromosome derived from the mother, this region binds an insulator protein that blocks the downstream enhancer from initiating transcription of *Igf2*. The same enhancer can therefore stimulate the transcription of *H19*. Conversely, on the paternally derived chromosome the insulator protein cannot bind to the methylated insulator region and the enhancer can act on *Igf2*. Several tissue-specific *differentially methylated regions* (DMRs) surround the two genes and ensure their appropriate tissue-specific expression. The imprinting mechanism for this locus is conserved between humans and Drosophila. Generally, imprinting in such diverse groups such as mammals, insects, and plants exploits conserved epigenetic silencing mechanisms that have developed in all eukaryotes, although quite distinct groups of genes can be imprinted in different species (Reik and Walter 2001a; Reik and Walter 2001b; Biliya and Bulla 2010).

5.3.2 Cytoplasmic inheritance

Cytoplasmic inheritance is important for both the broad and the narrow conception of epigenetics, Waddington's conception of the mechanisms that bridge genotype and phenotype, and the modern conception of heritable variations in gene expression. In the decades after the rise of Mendelian genetics many biologists continued to argue that additional hereditary factors are transmitted in the cytoplasm (Sapp 1987). Known forms of cytoplasmic inheritance now include the inheritance of organelles with their own genomes, such as mitochondria and chloroplasts, of internal and external membrane systems that are necessary as templates for the production of daughter membranes, and of maternal or paternal gene products such as transcription factors, mRNAs, and ncRNAs (which are part of narrowly defined epigenetic inheritance insofar as they establish self-sustaining feedback loops). All of these will influence the future development of the offspring.

Monozygotic twins derive from one zygote and therefore share all their genes. Nevertheless they develop differently. The nine-banded armadillo has been used extensively to study the development of monozygotic twins because this species always produces identical quadruplets. Experiments with armadillo have shown that not all four zygotes share the same intra-cellular environment. They all derive from the same fertilised egg in which the cytoplasm and intra-cellular proteins, mitochondria, and ribosomes are, as in all eggs, unequally distributed. This causes them to be unequally distributed to the resulting daughter cells and this in turn seems to contribute to the phenotypic differences between the armadillo siblings. Differences in the number of mitochondria can produce variations in energy production during development, and different gene products and chemical gradients in the egg cytoplasm can regulate early gene expression differently.

5.3.3 Regulatory RNAs

Not all RNAs are translated into proteins. RNAs can be processed into functional products such as structural RNAs or regulatory non-coding RNAs (ncRNAs). In recent years it has become evident that there are a large number of different regulatory RNA molecules which serve a wide range of functions (Morris 2008; Collins et al. 2011; Morris 2012).

Some ncRNAs are *trans*-acting factors that specifically direct epigenetic modifications, such as DNA methylation and histone modifications, to targets such as promoter regions and thereby modulate gene expression. This fact, plus the discovery of some unexpected modes of epigenetic inheritance associated with the cytoplasmic inheritance of RNA molecules, and the mitotic inheritance of differences in gene expression caused by ncRNAs, has led to RNA genetics often being discussed under the overall heading of epigenetics (Rassoulzadegan et al. 2006; Rechavi et al. 2011). For this reason it is discussed here rather than in Chapter 4. In that chapter we argued that it is one of the main goals of molecular biology to understand how genes are regulated during development and differentiation so as to ensure that genes are expressed in the correct time- and tissue-specific manner. We argued for the idea of a 'reactive genome' that is receptive to signals from outside the genome. Epigenetics, in the narrow sense, is one of the main fields that throws light on these issues, and non-coding, regulatory RNAs have emerged as one among other major players in regulating the expression of the genome.

The first proposed function for RNA was as the messenger from DNA to proteins (mRNA). Key roles were also found for RNAs in translation, such as transfer RNAs (tRNA) which link codons to amino acids, and the ribosomal RNA (rRNA) that forms the ribosomes together with small protein particles. It was later recognised that RNAs can fulfil catalytic functions, something previously reserved for protein-based enzymes. Ribozymes, for example, are the main enzymatic element in ribosomes (see Table 5.1a). In the 1990s it was realised that DNA is frequently transcribed into RNAs that neither code for polypeptides nor fulfil infrastructural or catalytic functions (Mattick 2001; Eddy 2001; Eddy 2002; Mattick 2003). Since then it seems that on a regular basis a new kind of regulatory RNA is described (see Table 5.1b and Table 5.1c).

One of the most important classes of small regulatory RNAs, micro RNA (miRNA), has been known since the early 1990s, but its regulatory significance only emerged in the early 2000s. All eukaryotic organisms use miRNA, and their highly conserved nature suggests that they represent an ancient form of gene regulation. In metazoans they are expressed ubiquitously in all organs and tissues after which they can 'circulate in a stable, and cell-free form in the bloodstream' (Kroh et al. 2010, 298). Micro RNAs usually induce gene silencing through partial base-pair complementarity with the target mRNAs to which they bind. Targeted mRNA in animals is normally not degraded, as happens in plants, but just prevented from being translated. Micro RNAs form a complex regulatory network because individual miRNAs may target a wide range of different mRNAs which in turn may contain multiple binding sites for different miRNAs (Hüttenhofer et al. 2001; Robinson 2009).

A closely related regulatory mechanism with many similarities to miRNA is RNA interference (RNAi) which operates by means of small interfering RNAs (siRNAs) that, as their name suggests, interfere with the expression of targeted genes. Like miRNAs they are short RNAs of circa 20–25 nucleotide length. Small

Table 5.1a *A list of non-coding RNAs with various functions: infrastructural and catalytic RNA*

Name	Function
Infrastructural RNA	
Transfer RNA (tRNA)	Translation
Ribosomal RNA (rRNA)	Translation
Signal recognition particle RNA (7SL or SRP RNA)	Membrane integration
Catalytic RNA	
RNA enzymes (Ribozymes)	RNA molecule with a well-defined tertiary structure that enables it to catalyse a chemical reaction
Group I / II self-splicing introns	Splicing
Riboswitches (aptamers)	Self-regulatory mRNA that binds target metabolites and thereby changes their secondary structure in response
RNA thermometer	Reacts with secondary structure to temperature fluctuations and activates heat shock transcription factor 1
Other housekeeping (infrastructural and regulatory) RNAs	
Small nuclear RNA (snRNA) + snRNP = spliceosome	Splicing
Small nucleolar RNA (snoRNA) + snoRNPs = editome	rRNA modification; possible post-transcriptional regulation
Cajal body-specific RNA (scaRNA)	RNA modification
Guide RNA (gRNA)	Editing
SmY RNA (smY)	Nematode *trans*-splicing

interfering RNAs normally function through complete sequence complementation, and promote cleavage of the targeted mRNA. Hence both miRNAs and siRNAs function through the silencing and/or degradation of targeted gene products. Both have also been implicated in the regulation of gene expression by directing chromatin remodelling (RNAi also provides a potent antiviral mechanism by targeting viral RNA).

Table 5.1b *A list of non-coding RNAs with various functions: short regulatory RNAs*

Name	Function
Small interfering RNA (siRNA/ RNAi)	Degradation, gene regulation
MicroRNA (miRNA)	Silencing, ca. 1,200 in human genome regulating 1/3 of protein-coding genes
Cis and *trans* antisense RNA (asRNA)	Degradation/silencing, gene regulation
Double-stranded RNA (dsRNA)	Gene regulation
Piwi-protein interacting RNA (piRNA)	Genome defence, heterochromatin formation, 50,000 alone in mice
Repeat associated small interfering RNA (rasiRNA)	Subclass of piRNAs
Transcription initiation RNA (tiRNA)	Tiny RNAs associated with TSS

A more recently discovered class of ncRNAs is long non-coding RNAs (lncR-NAs). It appears that these function both through *cis-* and *trans-*regulators of protein-coding gene expression: for example, through the activation of transcription factors or the modification of chromatin structure. They also seem to be involved in X chromosome inactivation, heat shock response, cellular immune response, and in nuclear architecture and neural functions (see Table 5.1c) (Lipovich et al. 2010; Chen and Carmichael 2010).

One of the early pioneers of Rnomics, the study of regulatory ncRNAs, John Mattick, has argued that regulatory ncRNAs comprise a whole new layer of complexity in the regulation of gene expression (Pang et al. 2005). He hypothesised that

> the principal advance in complex organisms was the development of a *digital programming system* based on ncRNA signaling, which bypassed the complexity limits that are imposed by accelerating regulatory networks that operate with proteins alone. (Mattick 2004, 317, italics added)

According to Mattick the main problem in the evolution of multi-cellularity was how to create *ordered* complexity. Just increasing the combinatorics of a higher number of interactions between *cis*-acting sequences and *trans*-acting proteins, such as transcription or splicing factors, must have early on reached a natural limit because of the 'nonlinear relationship between regulation and function' in functionally integrated systems (318). Each new structural protein necessitates several new regulatory proteins, each of those regulatory proteins

Table 5.1c *A list of non-coding RNAs with various functions: large non-coding RNAs (LncRNA)*

Name	Function
Xist	X chromosome inactivation
HOX antisense intergenic RNA (HOTAIR)	Controls gene expression on chromosome 2 by regulating its chromatin state
Human accelerated region 1 (HAR1F)	Active in the developing human brain between the 7th and 18th gestational weeks
CRISPRs (clustered regularly interspaced short palindromic repeats)	Prokaryotic immune system that confers resistance to exogenous genetic elements
Large intervening non-coding RNAs (lincRNA)	Gene regulation
Short interspersed nuclear (SINE)/Alu elements in humans	Self-propagating retrotransposons; gene regulation (splicing, editing, silencing)
Heat shock RNA-1 (HSR-1)	Temperature-sensitive non-coding RNA involved in mammalian heat shock response
Paraspeckles	Nuclear reserves of lncRNAs that activate transcription of stress-response genes
General: four times as common as mRNAs; lineage specific, low conservation with strongly conserved regions, incl. conserved secondary structures	E.g.: enhancer of TF functions, regulation of transcription machinery (control of promoter usage, inhibition of active initiation complex), histone modification

being coded by a sequence of thousands of nucleotides. That is several orders of magnitude more than needed to code for most ncRNAs. Mattick uses the terms 'analogue' and 'digital' to distinguish between regulatory molecules whose specificity depends on their three-dimensional shape and stereochemical specificity (analogue), and regulatory molecules whose specificity depends on the linear correspondence between nucleotide sequences (digital). The evolution of the ncRNA digital control system amounts to a transition to a more powerful genetic 'operating system'.

5.3.4 Transposable elements

A large proportion of eukaryotic DNA consists of transposable elements (TEs). These mobile genetic elements make up about 50 per cent of the mammalian, including human, genome. They copy themselves seemingly randomly into new places of the genome, which has earned them the name 'selfish' or 'parasitic' DNA (Doolittle and Sapienza 1980; Orgel and Crick 1980). Some evidence, however, suggests that whatever their origin they have evolved to become a part of the genomic community and have taken on new regulatory functions, either through the provision of novel coding and regulatory sequences for functional genes (particularly promoters and enhancers), or as novel ncRNA transcripts, since several of these functional elements in the genome seem to bear signs of having derived from TEs (Thornburg et al. 2006). For example, stress-induced expression of Alu and SINE 1 elements and their repression of Pol II transcription seem to have been recruited by the cellular stress response (Häsler et al. 2007; Walters et al. 2009; Ponicsan et al. 2010). A heated debate surrounds the question of whether the majority of TEs should be regarded as junk or rather as (newly recruited) functional elements.

5.3.5 Epigenetics in medicine

An increasing body of evidence supports the role of epigenetics in disease susceptibility. This research suggests that interaction with the environment may exert a major influence on health and disease in humans mediated by epigenetic mechanisms of gene expression.

According to the *Fetal Programming Hypothesis* the human fetus reacts with vascular, metabolic, and endocrine adaptations to circumstances in its environment. It is thought that nutritional or hormonal factors in the intrauterine environment induce epigenetic changes which affect the trajectory of prenatal development (Nathanielsz and Thornburg 2003). It has been argued such programming represents a form of *adaptive developmental plasticity* (see 5.5) which allows organisms to adapt to a suite of different environments with the most suitable phenotypic variant (Gluckman and Hanson 2005a; Gluckman and Hanson 2005b; Gluckman et al. 2007; Gluckman et al. 2009; Bateson et al. 2004; Bateson and Gluckman 2011).

In addition to epidemiological evidence and research on animal systems, molecular evidence has emerged in the past few years that underpins the proposed causal relationship between the fetal environment (e.g. maternal nutritional state or stress) and permanent changes in adult morphology, physiology, and behaviour (e.g. diseases such as obesity and metabolic syndrome or depression). Human infants who are exposed to under- or malnutrition in the womb and then encounter an abundance of food later in life develop obesity, cardiovascular disease, and other problems because the current environment was not predicted by the uterine environment (Godfrey et al. 2011). It has been proposed that the observed plasticity in human (and nonhuman) developmental trajectories is achieved through the altered expression of key regulatory genes that regulate cell number and differentiation early in development, and which can permanently reset the levels of activity of many physiological homeostatic mechanisms. These epigenetic processes are induced by environmental cues mediated by the placenta. It has been shown that particular maternally imprinted genes are targeted in fetal programming through the omission of epigenetic marks in certain tissues. Other genes which are normally not imprinted, however, can also be the target of selective activation or silencing (for a review of this work, see O'Malley and Stotz 2011).

The interpretation of such effects as adaptive developmental plasticity has led to them being labelled 'predictive adaptive responses' (PARs). PARs are designed responses to environmental cues that shift developmental pathways to modify the phenotype in expectation of a particular later environment. These changes may only manifest their adaptive effect later in life. The advantage of such a plastic strategy crucially depends on the accuracy of the forecast of the postnatal environment (Gluckman et al. 2005). A thrifty phenotype with a high ratio of fat cells versus muscle cells, a highly efficient metabolism designed to make the most of a meal, and changed appetite and exercise regulation may have clear advantages in an environment with poor nutritional supply, but would likely lead to highly increased weight gain and an increased risk of associated diseases in an environment with an overabundance of high-fat food. Such a scenario has been dubbed the 'Environmental Mismatch Hypothesis' (Gluckman and Hanson 2006).

We have introduced this research under the heading of epigenetics, but much of it has been pursued under other research agendas. In the next two sections we will introduce other approaches relevant to understanding the role of the environment in the regulation of the genome. The sets of biological phenomena which are studied under these agendas overlap, and all of them overlap with epigenetics. In 5.6 we will suggest an integrative framework.

5.4 Parental effects

Among the most important of these research agendas is work on 'parental effects'. As its name suggests work of this sort does not start from findings about underlying mechanisms. Instead, it starts from the relationship between parent and offspring phenotypes. Parental effects are sustained influences on offspring phenotype that are derived from the parental phenotype and are independent of the nuclear genes inherited by the offspring. More formally, we can say that a parental effect is a correlation between offspring and parent phenotypes which is not accounted for by either the correlation between their genotypes or the correlation between their environments.

Many parental effects connect the environmental experiences of the parental generation to the phenotype of the offspring. These are sometimes called 'environmentally induced parental effects' (Lacey 1998). In locusts, an environment overcrowded with conspecifics experienced by the mother causes her to coat her eggs with a hormonal substance. This substance contains serotonin which induces the egg to develop into a high-density morph with wings and legs suitable for migration (Anstey et al. 2009). Many parental effects, like this one, enhance the offspring's fitness. Natural selection has shaped offspring to respond to subtle variations in parental behaviours or parental provisioning as a forecast of the environmental conditions they will ultimately face after independence from the parent (Mousseau and Fox 1998; Weaver et al. 2004; Uller 2008; Maestripieri and Mateo 2009). The organism's developmental plasticity utilises environmental cues or developmental resources inherited from the parents to fine-tune its phenotype to the current or expected environment (Angers et al. 2010).

Because parental effects are defined phenomenologically, as an observable relationship between phenotypes, any mechanism which produces this relationship counts as a parental effect. The domain of phenomena that count as parental effects includes narrow-sense epigenetic effects that are reproduced in meiosis and thus can pass from one generation to another, but it includes many other things as well. The mechanisms that can create a parental effect include:

parental gene products (mRNAs, ncRNAs, proteins);

cytoplasmic inheritance (mitochondria, plastids, membranes, signalling factors, chemical gradients, intra-cellular symbionts);

oviposition (the placement of eggs in insects, fish, and reptiles can affect food availability and quality, temperature and light conditions, and protection against predators and other adverse conditions, and hence has important consequences for the fitness of the offspring);

gut organisms (which are often necessary for the normal development of intestines and the immune system, and daily metabolism);

sex determination (via maternal influence on temperature exposure in reptiles, hormonal influence on gamete selection in birds);

nutritional provisioning (prenatally through seeds, eggs or placenta, postnatal feeding particularly in mammals and birds, which is not just sustenance for the offspring but influences later food preferences and feeding behaviour);

parental care and rearing practices (warmth, protection, emotional attachment – e.g. differential licking in rats – and teaching and learning);

social status (in hierarchically organised mammals, such as primates, offspring often inherit the social status of the mother). (Mousseau and Fox 1998; Maestripieri and Mateo 2009)

Although most of these phenomena do not count as narrow-sense epigenetic inheritance, because they do not involve the transfer of chromatin modifications through meiosis, the phrase 'epigenetic inheritance' is sometimes used in a wide sense that is more or less equivalent to parental effects. We prefer to use the less ambiguous phrase 'exogenetic inheritance' in this context.

As might be expected from such a diverse field, there are many different approaches to parental effects. Parental effects researchers Alexander Badyaev and Tobias Uller (2009) have shown how the differences in the ways parental effects are understood reflect the different roles they play in research. These different approaches do not necessarily count exactly the same phenomena as parental effects. For many geneticists parental effect is essentially a statistical concept. It is an additional parent–offspring correlation (or anti-correlation) which must be added to a quantitative genetic model in order to correctly predict the effects of selection. In contrast, someone studying animal development is likely to define parental effects at a mechanistic level, referring to specific ways in which they are produced. Evolutionary biologists see parental effects either as adaptations for phenotypic plasticity, or as the consequence of a conflict between parent and offspring seeking to influence the offspring phenotype to suit their own interests.

> [P]arental effects mean different things to different biologists – from developmental induction of novel phenotypic variation to an evolved adaptation, and from epigenetic transference of essential developmental resources to a stage of inheritance and ecological succession. (Badyaev and Uller 2009, 1169)

We suggest that the distinctive feature of parental effects is that it is a phenomenological concept. So parental effects should not be defined by any specific mechanism that brings them about. Second, we would argue that parental effects should not be defined as adaptations. The evolutionary significance of parental effects does not depend on this – the correlations have the same impact on the dynamics of evolution whether or not they are adaptations! From a developmental perspective, parental effects need to be understood before the difficult question of their evolutionary origins can be properly addressed.

Badyaev and Uller recognise a point that we have tried to stress throughout Chapter 4 and this chapter: non-genetically inherited resources shouldn't be understood as competing with genetic resources; they complement them by amplifying the sequence information encoded by nucleic acids. We quote Badyaev and Uller's summary of the significance of parental effects in full because it touches on several points that are central to this and the preceding chapter:

> Here, we suggest that by emphasizing the complexity of causes and influences in developmental systems and by making explicit the links between development, natural selection and inheritance, the study of parental effects enables deeper understanding of developmental dynamics of life cycles and provides a unique opportunity to explicitly integrate development and evolution. We highlight these perspectives by placing parental effects in a wider evolutionary framework and suggest that far from being only an evolved static outcome of natural selection, a distinct channel of transmission between parents and offspring, or a statistical abstraction, parental effects on development enable evolution by natural selection by reliably transferring developmental resources needed to reconstruct, maintain and modify

genetically inherited components of the phenotype. The view of parental effects as an essential and dynamic part of an evolutionary continuum unifies mechanisms behind the origination, modification and historical persistence of organismal form and function, and thus brings us closer to a more realistic understanding of life's complexity and diversity. (Badyaev and Uller 2009, 1169)

In 5.6 we will suggest that the concept of the 'developmental niche' (West and King 1987) is the best framework to unite all the different approaches which study those aspects of heredity and development which are 'outside the genome'. Like Badyaev and Uller, we see the central issue as how *organisms reliably transfer developmental resources so as to construct and modify phenotypes in the next generation, primarily through regulated expression of the genome.*

5.5 Other ways to look 'outside the gene'

In this section we outline three other research agendas which have their own distinctive perspectives on how the environment, acting through the regulatory mechanisms described in Chapter 4, plays an instructive role in regulating gene expression.

5.5.1 Developmental plasticity

This field deals with the ability of organisms to react flexibly to different environmental conditions. Much research on developmental plasticity focuses on cases in which plasticity is adaptive and constitutes a means for organisms to match their phenotypes to spatially or temporally variable environments, as we saw in the discussion of medical epigenetics in 5.3.

The last century saw enormous progress in genetics. Despite the accepted truism that the phenotype results from the interaction of genes and environment the systematic investigation of environmental effects on the phenotype was neglected. It has been argued that in evolutionary biology the study of plasticity was stigmatised by the ghosts of Lamarck and Lysenko, but it was also lost from view in developmental biology. Embryology was less interested in variation within species than in fundamental characters shared by larger taxonomic groups (Amundson 2005, 183). Moreover, in the interests of replicability, developmental biology studied development using carefully monitored environmental conditions. Later the molecularisation of developmental biology made its study more dependent on model systems which were designed to minimise the effect of environmental noise on development (Bolker 1995).

> In 1965, when most geneticists considered developmental variability to be uninteresting noise, the genecologist A.D. Bradshaw coined the term 'phenotypic plasticity' to emphasize that environmentally contingent phenotypic expression could be a mode of individual adaptation to immediate challenges or stresses [...] Over the past two decades, plasticity has been studied intensively, with the primary goals of characterizing phenotypic variation expressed under diverse environmental conditions and assessing its potential evolutionary impact [...] This area of research has increased awareness that the context dependence of development is 'the rule rather than the exception' [...] and that it constitutes a fundamental mode of adaptive variation for many traits and taxa. (Sultan 2007, 575)

5.5.2 Ecological developmental biology

Ecological developmental biology (eco-devo) is the study of how development interacts with the ecological context in which organisms develop outside the laboratory (Gilbert 2001; Gilbert and Epel 2009). It is sometimes called 'eco-evo-devo' to identify it as an extension of evolutionary developmental biology, the successful new field that studies how development evolved, and how the ways in which organisms develop constrains and facilitates evolution. Ecological developmental biologists aim to understand not just how development interacts with ecology, but how the ecological aspects of development affect evolution. We will return briefly to these wider research agendas in Chapter 8.

> Ecological development or 'eco-devo' examines the mechanisms of developmental regulation in real-world environments, providing an integrated approach for investigating both plastic and canalized aspects of phenotypic expression. This synthetic discipline brings a current understanding of environmentally mediated regulatory systems to studies of genetic variation, ecological function and evolutionary change [...] Eco-devo is not simply a repackaging of plasticity studies under a new name but a more inclusive conceptual framework for understanding development in general [...] Whereas plasticity studies draw on quantitative genetic and phenotypic selection analyses to examine developmental outcomes and their evolution as

adaptive traits, eco-devo adds an explicit focus on the molecular and cellular mechanisms of environmental perception and gene regulation underlying these responses, how these signaling pathways operate in genetically and/or ecologically distinct individuals, populations, communities and taxa. (Sultan 2007, 575)

5.5.3 Epigenetic (exogenetic) inheritance

The idea of epigenetic inheritance was propelled to prominence by Eva Jablonka and Marion Lamb with their provocative title *Epigenetic Inheritance and Evolution: The Lamarckian Dimension* (Jablonka and Lamb 1995; see also Badyaev and Uller 2009; Bondurianasky 2012). Epigenetic inheritance in this context is used primarily in its narrow sense of the inheritance of chromatin modification, although as we will see below it often spills over into a broader sense. Jablonka and Lamb's identification of epigenetic inheritance with the inheritance of acquired characters is not unproblematic. Some scientists insist that the term 'Lamarckian inheritance' should be restricted to the inheritance of phenotypic (somatic) characters that are acquired during development (Hall 2011, 11).

Epigenetic inheritance differs in several important ways from nuclear genetic inheritance: epigenetic variations may be less stable than genetic ones, because these variations are in principle reversible, more sensitive to the environment, more directed, and more predictable, all features which may make them more adaptive in the short term than blind genetic variation (Jablonka and Lamb 1995; Holliday 2006, 78f.).

Some molecular biologists have argued that one should speak of epigenetic inheritance in the literal sense only in those cases when the methylation pattern is transmitted unchanged over several generations:

> [I]f epimutations is to have evolutionary importance, it must persist [...] This matter is central to whether epimutations can be treated as equivalent to conventional mutations or whether, if they have some degree of stability, some new population genetic theory is needed. (Wilkins 2011, 391)

Some cases certainly meet this criterion. In a comprehensive review of transgenerational epigenetic inheritance Jablonka and Raz (2009) conclude that epigenetic inheritance is ubiquitous, and can show stability of transmission of up to three generations in humans and up to eight generations in other taxa. However, many cases would not meet the criterion of multi-generational transmission. Epigenetic signals are very sensitive to environmental factors in that they are first 'established by transiently expressed or transiently activated factors that respond to environmental stimuli, developmental cues, or internal events' (Bonasio et al. 2010, 613). Many hypotheses about the evolutionary origins of epigenetic inheritance stress its value in spatially and temporally heterogeneous environments, where it allows rapid responses to change. However, the criterion of multi-generational stability may not be one we should accept. It is simply not correct that epigenetic change will only affect evolution if the changes themselves persist for more than one generation. Parental effects researchers have long known that one-generation parental effects substantially alter the dynamics of evolutionary models, and change which state a population will evolve to as an equilibrium (Lande and Price 1989; Wade 1998). In conventional quantitative genetics, the importance of Mendelism is not that individual genes can be tracked from one generation to the next – quantitative genetics does not do this – but that Mendelian assumptions let us work out what phenotypes (and hence fitnesses) will appear in the next generation as a function of the phenotypes in the previous generation. Epigenetic inheritance changes that mapping from parent to offspring, and this will affect evolution. There is no more central instance of the study of heredity than quantitative genetics, so more argument is needed for why epigenetic inheritance needs to be stable for several generations to be regarded as a form of heredity.

As we mentioned above, discussion of epigenetic inheritance often spills over from discussion of the specific phenomena of meiotic inheritance of chromatin modifications to include other phenomena which produce a parental effect. This is understandable, because narrow-sense epigenetic mechanisms are often important in parental effects which do not involve actual epigenetic inheritance. For example, in one well-studied example, epigenetic mechanisms have been shown to mediate the transgenerational effect of maternal care in rats without actual epigenetic inheritance. Maternal behaviour establishes stable patterns of methylation in the pups. These patterns affect brain development and the behaviour of the next generation of mother rats. The behaviour of those mothers re-establishes the patterns of methylation. But the actual patterns of methylation are not inherited (Meaney 2001a; Meaney 2001b; Meaney 2004; Champagne and Curley 2009a; Champagne and Curley 2009b). So long as the environment is constant, the phenotype will remain constant. If it changes, so that mother rats change their behaviour, the pheno-type will change with a one-generation time lag. Meaney and Szyf (2005) call this environmental programming of certain types of behaviour through DNA methylation 'life at the interface between a dynamic environment and a fixed genome'.

In a recent book, Jablonka and Lamb have attempted to organise the topic of epigenetic inheritance in this wider sense around four 'dimensions' of hered-ity: genetic, epigenetic, behavioural, and symbolic (Jablonka and Lamb 2005). The Genetic Inheritance System comprises protein-coding and non-coding RNA genes plus the regulatory motifs in the genome, as well as sequences with unknown functions. The Epigenetic Inheritance System includes modi-fications of DNA and chromatin, which are part of the nucleus. Besides these resources that are literally physically attached to the genome other devel-opmental resources are transmitted through the cytoplasm of the egg, such as parental gene products. The cortical (cytoplasmic) inheritance system, a subset of the overall epigenetic system, consists of cellular structures such as organelles with their own membranes and genes (mitochondria and chloro-plasts), membrane-free organelles (ribosomes and the Golgi apparatus), and the cellular membrane systems. All these structures cannot be produced from genetic information but act as templates for themselves. A Behavioural Inher-itance System (and we would add Ecological) forms a third dimension, in which information is transmitted through behaviour-influencing substances, non-imitative and imitative social learning, as well as habitat construction, food provisioning, and other parental effects like that described in the pre-vious paragraph. The last dimension is formed by the Symbolic Inheritance System (and we would add Cultural, although that could have been added to the Behavioural Inheritance System with equal justification, as Jablonka and Lamb seem to suggest). Offspring inherit social structures and rules, cultural traditions and institutions, and technologies. Importantly, this inheritance system includes epistemic tools, such as language, competent adults, teaching techniques, and so on. All systems use different mechanisms of transmission and show changing degrees of fidelity. Some mechanisms may not be intrin-sically stable. The Genetic Inheritance System, for example, relies on several layers of proofreading and copy-error detection systems for its exceptionally high fidelity. A suitable mechanism of scaffolding can lend the transmission mechanism reliability: proofreading supports genetic inheritance; epigenet-ics stabilises gene expression. Learning is scaffolded by teaching or by the reliable affordances of stimuli 'that define what is available to be learned [and] function to channel malleability into stable trajectories' (West et al. 2003, 618).

We have reviewed five agendas under which scientists have pursued research into what lies 'outside the genome': epigenetics, parental effects, developmental plasticity, eco-devo, and epigenetic inheritance. In the next section we will try to integrate these approaches using the idea of the developmental niche.

5.6 The developmental niche

> Natural selection acts to select genomes that, in a normal developmental environment, will guide development into organisms with the relevant adaptive characteristics. But the path of development from the zygote stage to the phenotypic adult is devious, and includes many developmental processes, including, in some cases, various aspects of experience. (Lehrman 1970, 36)

The concept of the ontogenetic, or developmental, niche was introduced in 1987 by developmentalist psychobiologists Meredith West and Andrew King (1987). It provides a way to bring together the research agendas we have described above, while highlighting just those features of each that are most relevant to the argument we have been developing. We have argued that biological specificity is distributed across the genome and its regulatory mechanisms, and that those regulatory mechanisms involve many factors 'outside the genome', including aspects of the environment and of experience, as highlighted by Lehrman above. These factors are not merely permissive, but instructive. They act as causally specific difference makers for the regulation of genome expression. Many aspects of the environment and experience of a developing organism are there by design: 'genes inherit a rich and supportive environment, a fact few dispute but few discuss with any urgency' (West and King 1987, 552). Evolution has designed not only a reactive genome, but also a developmental niche that reacts with it to construct phenotypes.

West and King (1987, 550) define the ontogenetic niche as a 'set of ecological and social circumstances inherited by organisms'. The developing organism can expect to encounter this niche in development as reliably as it does its genome: 'It's the dependability of the niche in delivering certain resources to the young that makes it a legacy' (West et al. 1988, 46). We would add epigenetic, epistemic, cultural, and symbolic legacies to this list and point to Jablonka and Lamb's (2005) 'dimensions' of heredity as a thorough and principled effort to taxonomise the contents of the developmental niche. Naturally, some dimensions are more prominent in one taxon than another. Together, these legacies are designed to provide the developmental resources needed to reconstruct the life cycle in each generation.

The developmental niche provides an alternative to the nature/nurture dichotomy (West and King 2008; Stotz 2008; Stotz and Allen 2008). The niche equals nurture, since it nurtures the developing organism, and it equals nature because it is part of the organism's endowment. West and King and their collaborators devoted decades of painstaking research to the acquisition of species-typical behaviour by the Brown-headed Cowbird. The cowbird is a nest parasite, and it had been assumed that, since it could not learn species-specific behaviour from its parents, those behaviours must be 'innate' or 'in the genes'. The cowbird was Ernst Mayr's flagship example of genetically encoded behaviour (Mayr 1961). West and King set out to show that this kind of dichotomous thinking was no substitute for causal analysis of how the phenotypes actually develop. The results of this research, which are hard to assimilate to either side of the nature/nurture dichotomy, led them to develop the niche concept. The ability of cowbirds to recognise their own species visually depends, among other factors, on 'phenotype matching' – individuals learn what they themselves look like and then seek to interact with birds that look like that (Hauber et al. 2000). Male song is shaped by feedback from female cowbirds, whose wing stroking and gaping displays in response to the songs are strong reinforcers for males (West and King 1988). Raised in isolation, males will sing, but they need feedback from a female audience and also competition with other males in order to learn how to produce cowbird songs in a way that leads to successful mating:

> In cowbirds the juvenile niche is a forum in which males learn the pragmatics of singing, which appears to be a performatory, if not sometimes martial, art. (West et al. 1988, 52)

Female song preferences are themselves socially transmitted (West et al. 2006). As a result, cowbirds reliably transmit not only species-typical song, but also regional song dialects. Ecological factors help ensure that cowbirds find themselves in flocks which make these various processes possible. The flock functions as an information centre controlling what is 'bioavailable' to be

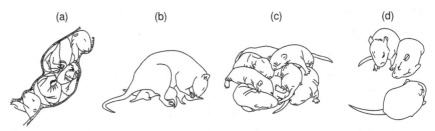

Figure 5.2 Four ontogenetic niches of rat ontogenesis: (a) the uterine niche with three adjacent fetuses; (b) the dam as niche, shown in nursing pose; (c) the huddle niche of some week-old pup siblings; (d) the coterie niche of older, more exploratory siblings. (Image courtesy of Jeffrey R. Alberts, Indiana University, www.indiana.edu/~ablab/).

learnt throughout the lifespan. The developmental niche concept undermines the traditional dichotomy between heredity and individual experience, since it highlights how experience, including in some taxa learning, are involved in the development of species-typical behaviour. Aspects of experience are part of the mechanism of heredity (West and King 2008).

The cowbird is not an isolated example. Other examples in which developmental niches afford the robust experiences necessary for normal development include food and habitat imprinting in insects through oviposition; maternal care and stimulation for neural development (sexual behaviour and fear reaction in rats; learning disposition in chickens); territorial and habitat inheritance (nest sites, food resources, a hierarchy of relatives) in woodpeckers and jays; maternal rank inheritance in carnivores and primates (Maestripieri and Mateo 2009).

Jeff Alberts has used the developmental niche extensively in studies of rat development. The rat pup passes through four consecutive 'nurturant niches' on the way to adulthood: the uterine niche, the dam's body, the huddle in the natal nest, and the coterie (Alberts 1994) (see Figure 5.2). They all provide sustenance for the developing organism, such as nutrients, warmth, insulation, and 'nurture' in the form of behavioural and social stimuli as affordances for development. The early ontogeny of species-typical rat behaviour is directed mainly by olfactory, but also tactile, cues that are provided by the different ontogenetic niches. Olfactory cues on the dam's nipples guide the pup to them. However, the pup's developing sensoria need to acquire odour recognition of the nipple through chemical cues in the amniotic fluid provided by the uterine niche it has passed through before. The spread of amniotic fluid over the dam's body after birth bridges the pre- and postnatal niches of the pup. Filial huddling preferences in the natal niche are mediated by learnt olfactory cues provided from the close proximity of the siblings during the suckling stage. This huddle or natal niche in turn induces preferences prerequisite for the functioning of the rat in the social context of the 'coterie niche', through thermotactile stimulation. Alberts notes:

> Again we find a stereotyped, species-typical, developmentally-fixed behavior is learned, with all of the key components [...] existing as natural features of the ontogenetic niche [...] Specific features of these niches elicit specific reactions and responses in the developing offspring. (Alberts 2008, 300)

These niches afford the pups a range of other experiences. In 5.5.3 we encountered Michael Meaney and collaborators' discovery that natural variation in maternal care, elicited by experiences of the mother, influence stress responses and exploratory and maternal care behaviour in the offspring. The quality of the mother's licking and grooming behaviour results in a cascade of neuroendocrine and epigenetic mechanisms. One pertinent example is the permanent down-regulation in the expression of the glucocorticoid receptor gene in the pup brain's hippocampus via the methylation of its promoters, which occurs in response to a low level of licking and grooming by the mother (Meaney 2001b; Champagne and Curley 2009a; Champagne and Curley 2009b). This down-regulation causes high stress-reactivity in the offspring. Hence stressed mothers in reaction to an adverse environment produce stressed daughters who in turn become stressed mothers. This is not necessarily bad, since highly stressed individuals are better prepared to survive in adverse environments (e.g. with a high level of predation). Conversely, relaxed mothers that show a high level of licking and grooming produce relaxed offspring that develop into high licking mothers (see Figure 5.3). This work demonstrates another way that the developmental niche concept makes some experiences part of the mechanisms of inheritance. Experience can help to construct the legacy that the next generation will receive: 'Exogenetic legacies are inherited, but they are also learned' (West et al. 1988, 50).

The developmental niche explains the reliable development of species-typical features, but the framework is equally applicable to plastic phenotypes. Many developmental systems are 'designed to be as open as ecologically possible and thus immediately sensitive to ecological change' (West and King 2008, 393). The niche contains the scaffolding for normal development, but

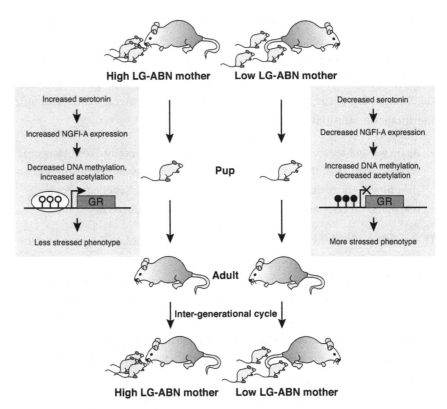

Youngson NA, Whitelaw E. 2008.
Annu. Rev. Genomics Hum. Genet. 9:233–57

Figure 5.3 The transgenerational transmission of stress-reactivity in rats. The figure shows the relationship between the dam's maternal care style (high or low licking and grooming with arched-back nursing), and epigenetic changes in gene expression in the hippocampus of the pup, leading to different maternal care styles in the next generation. (Printed with permission from Annual Reviews: Neil A. Youngston and Emma Whitelaw, 2008. Transgenerational epigenetic effects. *Annual Review of Genomics and Human Genetics* 9: 233–57.)

the genome has coevolved with the niche and can also use it as a source of information for developmentally plastic responses: 'Animals have evolved to integrate signals from the environment into their normal developmental trajectory' (Gilbert and Epel 2009, 9). The fact that development is not laid out before it occurs, with other causal factors as merely permissive (or disruptive), but instead emerges through a process of epigenesis is what enables the integration of robustness and plasticity in development (Lamm and Jablonka 2008; Bateson and Gluckman 2011).

5.6.1 Niche construction

Along with the idea of the developmental niche goes the idea of developmental niche *construction*. Organisms place DNA into a developmental setting that is always highly characteristic of a lineage and which commonly owes much of its structure to the activity of the parental and earlier generations. The niche is constructed through those activities.

> The utter reliability of the ontogenetic niches and the affordances that exist in each are inherited as surely as are genes. An offspring's behavioral interactions with the dam or with its siblings in the nest can be framed as active 'niche construction'. (Alberts 2008, 301)

This raises the question of how developmental niche construction is related to the idea of 'niche construction' introduced by John Odling-Smee (1988), and developed by Odling-Smee, Kevin Laland, and Marcus Feldman (2003). Niche construction is an evolutionary concept which refers to the active construction by a population of a selective environment which in turn influences the selective pressure on that population. Niche construction builds on earlier suggestions that organisms are not passive in evolution, but actively construct their environments, which in turn influence the organisms (Lewontin 1982).

Many of the processes by which organisms influence their own environments are examples of both developmental and niche construction. For example, by nesting in a tree cavity a bird both constructs the developmental niche of its young and exposes them to different selection pressures, such as increased parasitism. Both ideas stress the active role of the organism in constructing aspects of the environment, and both explore the implications of non-genetic causal pathways between parents and offspring. Nevertheless, the developmental niche should not be conflated with the selective niche. The selective niche is a set of selection pressures and is normally studied at the level of the population and on evolutionary timescales. The developmental niche is a set of developmental resources and is normally studied at the level of the individual organism and on the timescale of individual development.

The fact that the two senses of 'niche construction' are conceptually distinct does not mean that endorsing one idea has no implications for the other, but those implications are complex. (For an exploration of the closely related issue of how niche construction impacts on ecological developmental biology and vice versa, see Laland et al. 2008.)

5.7 Conclusion

This chapter continues the argument started in Chapter 4 against the conventional idea that hereditary information is encoded in DNA. In that chapter we saw that regulated expression of the coding regions of the genome depends on regulatory mechanisms that differentially activate and select the Crick information in coding sequences. We argued that in so doing these mechanisms create additional Crick information, contrary to an exclusive reading of Crick's sequence hypothesis. We also described some mechanisms that straightforwardly create additional Crick information by determining the order of some of the elements in gene products without reference to an underlying coding sequence. Sequence specificity, we argued, is distributed between the coding regions in the genome and regulatory mechanisms, and the specificity manifested in gene products is the result of a process of molecular epigenesis.

But this did not rule out the suggestion that all specificity traces back to some feature or other of the whole genome sequence. In this chapter we have tried to undermine that idea by showing how the environment, acting through the regulatory mechanisms described in Chapter 4, plays an instructive role in regulating gene expression. In 5.3 we described the molecular mechanisms of epigenetics, some of which are also mechanisms of epigenetic inheritance in the narrow sense of chromatin modifications that are reproduced in meiosis. Epigenetic inheritance can clearly act as an instructive cause in gene expression: it can turn genes on, up, down or off and affect splicing in a time- and tissue-specific pattern. In 5.4 we introduced the concept of parental effects, a phenomenological notion that applies to many different mechanisms that create correlations between parental and offspring phenotype independent of correlations between genotype. Once again, many parental effects have highly specific effects on development. So we conclude here that the specificity expressed in the time-and-tissue specific set of gene products produced by an organism cannot all be traced back to some feature or other of the whole genome sequence.

Some parental effects and some inheritable epigenetic marks (e.g. genomic imprinting) are produced by mechanisms in the parent that are under genetic control, in the sense that the specific actual difference makers (see 4.3) for the effect the parent has on the offspring are differences in DNA sequences in the parent. So perhaps all specificity traces back to some feature or other of *some* whole genome sequence, either that of the organism or that of one of its ancestors? We disagree. In 5.5 we introduced the idea of adaptive developmental plasticity. Organisms adjust their phenotypes in the light of information from the environment, often using some of the epigenetic mechanisms described in 5.3. Some parental effects are also instances of 'inter-generational adaptive developmental plasticity' – that is to say, the phenotype of the offspring represents a response to information from the environment received by the parent. Many forms of developmental plasticity are highly specific, so not all specificity traces back to some feature of the whole genome sequence of the organism or that of one of its ancestors.

In 5.6 we introduced the idea of the developmental niche and developmental niche construction. The developmental niche contains all the legacies that are required for properly regulated expression of the genome throughout the life cycle of the organism. We suggested that this is a good integrative framework for the various ideas about heredity 'outside the genome' introduced in earlier sections. Epigenetic inheritance in the narrow sense is a contribution to the developmental niche of the next generation by transmitting gene expression profiles. The diverse mechanisms that produce parental effects all contribute to developmental niche construction. 'Entrenched' parental effects contribute to the reconstruction of the developmental niche in each generation, while 'context-specific' parental effects modify the developmental niche in reaction to environmental parameters (Badyaev and Uller 2009, 1172). The latter are an example of inter-generational developmental plasticity, in which parents facultatively modify the developmental niche of their offspring. In both inter-generational and intra-generational developmental plasticity organisms use the developmental niche as a source of information about fluctuating environments. Finally, the concept of ecological developmental biology seems to be equivalent to the study of the developmental niche.

The genetic heredity system may well have been optimised for the ability of organisms to transfer sequence specificity between generations. Crucially, this ability is dependent on the invention of nucleic acid-based heredity. At least some exogenetic channels of inheritance may have been optimised for the ability to respond flexibly to and in concert with environmental demands. After all, epigenetic mechanisms react to environmental signals and allow for different levels of heritability, from stability throughout the lifetime of an individual, over a parent–offspring dyad, to long-term stability of a lineage. As Meaney and Szyf (2005) succinctly put it, epigenetics connects a fixed genome to a dynamic environment.

Further reading

Eva Jablonka and Marion Lamb have produced two excellent introductions to epigenetics and to epigenetic/exogenetic inheritance (Jablonka and Lamb 1995, 2005). Scott Gilbert and David Epel's *Ecological Developmental Biology* (2009) is a rich source of information on developmental plasticity, epigenetics, mutualistic relationships, and much more. Two important collections on parental effects are *Maternal Effects as Adaptations* (Mousseau and Fox 1998) and *Maternal Effects in Mammals* (Maestripieri and Mateo 2009). Patrick Bateson and Peter Gluckman (2011) have provided a very readable introduction to the interdependence of plasticity and robustness in development and evolution.

6   The gene as information

[G]enes are merely repositories of information written in a surprisingly
similar manner to the one that computer scientists have devised for the
storage and transmission of other information.

(Economist 1999, 97)

6.1 What everybody knows

The journalist who wrote those words would have felt on safe ground. Everyone knows that genes are composed of words written in the genetic code and
that together these make up a book, or an instruction manual, for the organism. In this information age one of the most prominent identities of the gene
is as information, code, programme, blueprint, recipe, and even 'book of life'.
These metaphors dominate popular understanding of genetics and molecular biology. When the Human Genome Project determined the sequence of
human DNA, the scientists were said to have decoded the book of life. The
efforts that followed to understand the functions of the many new and unexpected structures found in the genome and which we described in previous
chapters are apparently decoding the book of life a second time. If scientists
identify loci relevant to an illness, they have cracked the code for that disease.
When someone sequences those loci, they too will have cracked the code, or
perhaps deciphered the message, and the scientists who work out some of
the interactions between molecules that connect activity at that locus to the
disease will probably be congratulated on cracking the code for a third time.
It is easy to dismiss such clichéd and thoughtless use of information language.
However, many scientists are attached to the view that genes are, ultimately,
units of information. They see this as a profound insight into the nature of
living systems, and as one of the most important 'big picture' conclusions to
have emerged from genetics and molecular biology.

The influential evolutionary biologist George C. Williams proposed that genes are units of information, rather than physical objects made of DNA. Genetic information happens to be recorded in nucleotide bases, but this is comparable to the fact that this sentence is written in ink. The same sentence can be written in pixels (perhaps you are reading it that way), or in magnetic traces. Like the material form of the sentence, the material form of the gene is inessential. The gene itself is a unit of information. Williams' strongest statement of this view came at the end of his career (Williams 1992), but the idea of abstracting away from the physical nature of genes to focus on the information they contain was implicit in his earlier work (Williams 1966) and was popularised by Richard Dawkins (1976) with his idea of the gene as a 'replicator'. A replicator is something of which copies are made, such as a sequence of DNA or a book. Copies of a replicator may only exist for a short time, but a *kind* of replicator can potentially exist for ever, as long as copies of it persist. So if the replicator is the information that specifies the order of nucleotides in a piece of DNA, then we can say that replicators are 'potentially immortal', as Dawkins does. The molecule may pass away, but the information persists as long as any copy of it exists somewhere, in another DNA molecule, or elsewhere, perhaps in a computer database. It is even possible to claim that it is the information that is really evolving, rather than the physical objects that embody that information at each point in time.

The informational gene has given rise to heated debate within philosophy. There are profound disagreements about the role and value in biology of models drawn from language, communication, and computing. In part, this is because debates in the philosophy of mind and language have led philosophers to take substantial positions on the nature of information, and on what role information can play in scientific explanation. Molecular biology is a highly successful field of science that makes extensive use of information and related ideas. So it is natural that philosophers have looked to it for vindication of their more general views about information. But debates about information talk in biology also reflect differing views within philosophy of science about the nature of scientific representation and the relationship between models and the real-world systems that are their targets.

Before we begin to unpick these scientific and philosophical debates, two preliminaries are necessary. First, we need to distinguish two senses in which biologists talk of 'genetic information': information *in* genes and information *about* genes. We are only concerned with the first of these. Genetic information in this first sense is a theoretical entity which exists in the genome and explains biological phenomena. For example, the fact that DNA sequences contain information about the structure of proteins explains how cells possess and transmit biological specificity. But the phrase 'genetic information' is also used in a second sense: namely, information *about* genes. The majority of references to 'genetic information' in the philosophical literature use the phrase in this second sense. They occur in bioethical discussion of when it is permissible to obtain, store, share, or use information about someone's genome. Genetic information in this sense is important in biology too. Much of the work of 'bioinformatics' is about accessing, managing, manipulating, and representing the vast quantities of data produced by contemporary biology. In this second sense it is obvious that genomics – the sequencing, annotation, and functional analysis of entire genomes – is an information science. But so too is proteomics, the effort to characterise the entire protein complement of an organism and its behaviour, and phenomics, which attempts to relate the phenotypes of organisms to their genotype in an equally comprehensive manner. What makes biology an information science in this sense is not anything about the nature of genes, but the fact that contemporary biology works with vast bodies of data that the unaided human mind is incapable of processing effectively.

The fact that genetics and molecular biology are centrally concerned with genetic information in the second sense – information about genes – does not imply that biology is or should be centrally concerned with genetic information in the first sense – a theoretical entity found in genomes and which explain some or all of what genes are and what genes do. Additional arguments would be needed to show this. It is such arguments that we are concerned with in this chapter.

The second preliminary is a brief sketch of the historical emergence of the informational gene. In Chapter 3 we discussed how the stereochemical conception of biological specificity was supplemented by, and sometimes replaced by, an informational notion of specificity. This was only one aspect of a substantial change in the intellectual landscape of the biological sciences that started in the late 1940s and continued into the early 1960s. Genetics and molecular biology came to view their subject matter as the storage, transmission, and interpretation of biological information. By the time Watson and Crick proposed a structure for the DNA molecule in 1953 it was not surprising that they also suggested the order of bases in the molecule was a 'code' that could hold the 'biological information' of the genes. In 1958 Crick stated the Central Dogma not merely as a one-way causal process, but as a one-way flow of information from DNA to RNA to protein. The genetic code relating RNA codons and amino acids (Table 3.1) was worked out between 1961 and 1966. The idea that genes are to be understood as coded messages, that discoveries in molecular biology consist in decoding or deciphering the information in genes and other molecules, and the broader idea that the genome encodes a programme for the development of the organism, are now all commonplace. But these ideas were not prominent in either genetics or biochemistry until after the Second World War.

It is evident that the informational turn in genetics reflected the emergence of the information sciences: electronic computing machinery, the science of 'cybernetics', which grew out of the wartime study of self-directing machines, and the mathematical theory of communication, developed by Claude Shannon (Shannon and Weaver 1949). The complex interchange between the information sciences and biology is the subject of an extensive literature in the history of science (see Further Reading). While there are controversies about the exact nature of these influences, there is no doubt that mid-twentieth-century biology drew on these new technologies to interpret living systems. It was as inevitable that molecular biology would compare the genome to a computer tape as it was that earlier scientists would compare organisms to clockwork or to the steam engine. But the fact that scientific ideas are derived from the broader cultural context does not imply that they lack scientific justification. William Harvey compared the heart to a pump, a simple mechanical device of the kind that seventeenth-century scientists were inclined to see everywhere in nature. But in a very straightforward sense, the heart is a pump. Perhaps biology is equally straightforwardly an 'information science'.

6.2 Information, metaphors, and models

Philosophical debate over the significance of informational language in biology has centred on whether such language is literal or metaphorical. On the one side, we have the philosopher Alexander Rosenberg:

> [T]he genes literally program the construction of the Drosophila embryo in the way the software in a robot programs the welding of the chassis of an automobile. (Rosenberg, 2006, 61–2)

On the other side, we find the philosopher and biologist Sahotra Sarkar:

> [T]here is no clear, technical notion of 'information' in molecular biology. It is little more than a metaphor that masquerades as a theoretical concept [...] (Sarkar 1996, 187; see also Griffiths 2001)

But the contrast between metaphorical and literal is too crude to illuminate how informational language works in biology. It assumes, first of all, that the results of science *can* all be stated in a canonical language entirely devoid of metaphor. That assumption made sense in the first half of the twentieth century, when philosophers held that the primary products of science were theories and that theories could in principle be formalised as a set of axioms from which all the observational consequences of a theory could be deduced. The use of metaphors, similes, or analogies could only be a sign of scientific immaturity. The language of a mature scientific theory would use only observation terms defined by measurement procedures, and theoretical terms either defined by their role in the axioms of the theory or formally defined using existing terms. All relationships between terms would correspond to formal, logical procedures. There would be no more room for metaphor in science than in the axiomatised portions of mathematics, which mature scientific theories were thought to resemble.

One of the first philosophers who rejected this view was Mary Hesse, whose *Models and Analogies in Science* (1966) gave a systematic treatment of the role of analogy in the construction of scientific theories. She also demonstrated the historical importance of this process in science. For example, wave theories of light were constructed using an analogy with waves in physical media like water. Hesse distinguished three classes of analogy between the source of a metaphor and its target. Positive analogies are features of the source that are known to be shared by the target. Negative analogies are known not to be shared. Neutral analogies are those that have yet to be settled. Part of the heuristic role of metaphor and analogy in science is that they encourage scientists to investigate the neutral analogies between source and target.

Hesse (1988) used her case studies to argue that even in mature sciences the language in which the science is expressed remains metaphorical. But it has since become clear that the several ways in which science constructs 'models' of the world cannot be neatly classified into literal representations on the one hand, and metaphorical or analogical representations on the other. Since the 1960s the nature of scientific models, and modelling as a way of doing science, have become central topics in the philosophy of science. Faced with such a large literature all we can do here is to highlight one or two relevant, and we hope not too controversial, points about that general literature (for an introduction, see Frigg and Hartmann 2009) and then evaluate specific proposals about information-based models in biology.

The 'semantic view of theories' was first proposed in the 1960s and became highly influential in the 1980s (Suppes 1960; van Fraassen 1980). Some authors have argued that it is particularly plausible as an account of theory in biology (Lloyd 1988; Thompson 1989). On the semantic view, the linguistic or mathematical expression of a scientific theory serves to define an abstract structure. It is this structure which is the subject of the theory. This view of the subject matter of theories is familiar from mathematics. The subject of Euclidean geometry is Euclidean space, no matter whether the actual space in which we move is Euclidean. Euclidean space is a 'model' for Euclidean geometry in the mathematician's sense of the word 'model': it is something of which the statements of the theory are true. If there are no Euclidean spaces in the physical world, Euclidean geometry still remains the correct account of Euclidean space. The semantic view of theories treats scientific theories in the same way. The subject of Mendelian genetics is the class of systems which obey the principles of that theory. The abstract structure which characterises all such systems is a 'model' for Mendelian genetics in the mathematician's sense: the statements which Mendelian genetics says are true or false are true or false of that structure. A consequence of the semantic view is that theories represent actual physical systems only in an indirect way. If the inheritance of some trait does not obey Mendelian principles, this does not show that Mendelian genetics is false, but only that the mechanisms of heredity for this trait are not Mendelian. Rather than asking if a theory is true, the semantic view asks where the theory can be applied.

Classical statements of the semantic view analyse 'applicability' as the existence of an isomorphism between a real-world system and the structure specified by the theory. Is there a way to map parts of the abstract structure onto parts of the real-world system? Is there a way to map the relationships allowed by the abstract structure onto processes that occur in the real-world system? If so, then the theory can be applied to the real-world system: our understanding of what happens under certain conditions in the abstract structure can be used to predict and explain what will happen in the real-world system.

Recent authors have suggested a less formal, more psychologistic, way to think about scientific models, but one which has many of the same consequences. The simplified version of the world described in the model does not actually exist, so these authors compare the model to a work of fiction (Godfrey-Smith 2006; Frigg 2010; Levy 2011). When we reason about how things work in the model, this is like reasoning about a fictional world. Although the worlds described in works of fiction do not exist, there are still right and wrong answers about them. Sherlock Holmes is a fictional nineteenth-century British detective. Someone who has not read the books and thinks that Sherlock Holmes was an American detective in the 1970s is mistaken, even though Holmes does not exist. There are also subtler ways to be wrong about Sherlock Holmes. There is no Holmes story in which he meets Charles Darwin, but we can confidently say that if Holmes had met Darwin he would not have murdered him and pocketed the Darwin family silver. If you disagree, I could say 'You don't understand how his mind works.' Reasoning about models has something in common with reasoning about fictions. Although there are no perfectly rectangular cells, a scientist might try to work out how the receptors on the surfaces of cells interact in a model with rectangular cells. Another scientist could criticise the first scientist's reasoning and argue that something different actually happens in the model under those circumstances. The disagreement between these scientists is about the imaginary cells in the model, not about any actual cells. According to the fictionalist, models are useful because the imaginary world resembles the real world in key respects, so that we can reason from one to the other, but it is simpler than the real world, so we can more easily work out what happens in the imaginary world under some specified circumstances, and why it happens.

The early twentieth-century view of scientific theories as axiomatised deductive systems left no room for metaphorical language in mature science. One might say that the semantic view leaves no room for literal language, or at least literal language referring to real-world systems. Scientific theories are not about real-world systems. Instead, they are more or less applicable to real-world systems because they resemble those systems in various respects. But this is what a metaphor does: it draws our attention to the resemblance between two things and invites us to think about one by thinking about the other. So if the semantic view is correct, the way in which scientific theories relate to the actual world has much in common with the way in which metaphors relate to their targets. Fictionalism has a similar consequence: a fictional model is useful because it has points of comparison to the real world, which is just what makes metaphors and analogies useful.

The semantic view was intended to apply to all mature scientific theories, including theories whose scientific creators thought they were directly describing real-world systems. Since the 1980s philosophical discussion of models and modelling has become more concerned with actual scientific practice (Giere 1988) and has become sceptical of this one-size-fits-all approach to the structure of scientific theories (Downes 1992). For example, the original 1953 description of the structure of DNA by Watson and Crick was a model in several senses of that word. It was a model in the mathematician's sense: an object about which a set of statements was true, in this case a set of statements derived from empirical research about the structure of DNA. It was also a physical model – made with pieces of metal. More importantly, Watson and Crick were attempting to discover more about the DNA molecule by creating a structure known to resemble the molecule in certain respects, and then conducting research to discover if the actual molecule resembled the model in further respects. By examining and manipulating the model, they learnt about the structure of the actual molecule.

However, it is unclear that a description of the DNA molecule in a contemporary biochemistry text is a model in any of these senses. It seems more like a description of the arrangement of furniture in my living room: if we took a particular DNA molecule and compared it to the description, there would be an atom at each position where one is described, a positively charged hydrogen atom on one side of each of the bonds between the nucleotides mentioned in the description, and so forth. Biologists physically arrange atoms to make specific DNA molecules; they even patent some of the resulting arrangements. They use DNA molecules to make new kinds of organisms and, as Ian Hacking (1983) famously argued, it is hard not to believe in your own tools. Statements about the shape of the molecule are equally straightforward. For example, the double-helical form gives a mechanistic explanation of why, when crystallised DNA is bombarded with x-rays, they scatter in distinctive patterns. If the actual molecule does not have the shape of a double helix, this explanation does not work. It is still possible to apply the semantic view to this example: the description defines an abstract class of 'Watson–Crick structures'; even if DNA did not have this structure, it would still be true of its real subject, namely Watson–Crick structures; the referents of the terms in the description are not actual molecules, or bonds, but elements of an abstract structure which is isomorphic to the molecule. But this is a profoundly artificial exercise and does little to further our understanding of descriptions of DNA structure in molecular biology. The same could be said in response to the proposal that the structure of DNA is a fiction which has points of resemblance to the real world.

There is no need to force every instance of science into a single mould. Modelling can be seen as one way of doing science: namely, by adopting a strategy of 'indirect representation' (Godfrey-Smith 2006). Instead of studying a complex, real-world system, scientists elect to study a simpler system which resembles the target system in certain respects. The model system may be a specially developed actual system, like the laboratory strains of nematode worm used to study animal development at the molecular level, or a scale model in engineering. It may be a computer model, as in much of systems biology, or it may be purely imaginary, with no existence outside the scientific texts that discuss it. This strategy of indirect representation appears to be very common across a range of sciences from systems biology to climatology. It is an effective way to gain understanding of complex or inaccessible target systems in the real world.

If indirect representation is a legitimate and important form of scientific practice, then it becomes a question as to why calling a theory 'metaphorical' is a criticism. It seems the term 'metaphorical' is often used for a loose resemblance which fails to meet the standards required of a mature scientific model. Susan Haack had a similar thought when she criticised Hesse's claim that mature scientific theories are steeped in metaphor:

> Briefly and approximately, though, I see metaphors as, so to speak, rough drafts of scientific theories, rough drafts which offer guidance to possible directions for refinement and specification. (Haack 1987, 299)

A scientific metaphor, then, is a starting point for model construction. Points of resemblance between the target system and some other, hopefully better understood, system are the inspiration for one or more detailed models. But this model building need not result in a literal description of the target system. Even the mature theory may take the form of a model – an indirect representation which relates to the target system through a set of resemblances. If the research tradition develops a single, precise, comprehensive model, then the strategy of indirect representation can be abandoned and the content of the model can be regarded as a straightforward description of the target system. We can debate exactly how 'straightforward description' is to be understood, but we should not lose track of the difference between this outcome and another possible outcome of scientific research. This other outcome is that even the fully developed model has strengths and weaknesses. It may be good for one set of scientific purposes, but not all. It may be too abstract, or too detailed, or have a simplifying assumption that matters in one context but not in another. In this kind of situation more than one model is often used to illuminate different aspects of the system. If this is how things pan out, it will remain necessary to distinguish between the model and the target system. The question will be whether, given what the model is to be used for, it resembles the real-world system in the right way, or does so better than any other available model.

Many of the analogies between genetics and computing and information technology that are explored in this chapter can be understood as examples of this strategy of indirect representation: model building inspired by some metaphorical application of informational language. For example, we will see organisms described as 'reducing their uncertainty about the environment' by having one genome rather than another. This description highlights a way in which the elements of a heredity system can be mapped onto the mathematical theory of communication, a formalism originally intended to describe analogue telephone systems (Shannon and Weaver 1949). This gives rise to models in which heredity is indirectly represented as a signalling system connecting two or more generations in an evolving lineage. Ideas and results from information theory can be imported to enrich these models, and the lessons learnt from them can be used to understand actual heredity systems.

The diametrically opposed views of the value of information talk represented in the quotations at the beginning of this section can be better understood in the light of this discussion. Rosenberg thinks that representations of genomes in molecular developmental biology in which genes and the interactions between them are modelled as a Boolean network of switches which perform logical operations constitute a single, precise, comprehensive model which captures all the data about the functioning of the genome. He proposes to cease distinguishing between the model and the actual, molecular processes of development, and to treat the 'genetic programme' as a straightforward description of processes in the cell.

Sarkar, in contrast, draws attention to cases in which pursuing an informational metaphor led to a dead end (Sarkar 1996, 2005). Attempts in the late 1950s to deduce the genetic code as an elegant and efficient application of information theory to the known chemistry turned out to be entirely misguided, leaving the problem to be solved experimentally by biochemists. This was one of several projects inspired in this period by Crick's description of the relationship between DNA and protein as a flow of information. Sarkar argues that this metaphor did not develop into a single, precise, comprehensive model of the molecular mechanisms of the cell. Instead, there are a number of ways to use elements of information and computing theory to model different aspects of what goes on within and between cells. Strategies of indirect representation may still prove useful here, but the distinction between the models and the target system remains essential.

6.3 The genetic code

The most uncontroversial aspect of information talk in biology is the genetic code, described in detail in Table 3.1. Each three-nucleotide codon of messenger RNA corresponds to a specific amino acid. When RNA is translated into protein the sequence of codons determines the order of amino acids in a polypeptide chain, one or more of which makes up a protein. The linear sequence of amino acids is known as the 'primary structure' of the protein.

The genetic code encodes a very specific kind of information, which we have referred to as Crick information. Crick information is the specification of the order of amino acids in a polypeptide chain. It is the form of information that is distinctive of coding regions of DNA, although, as we saw in Chapter 4 and Chapter 5, later developments in molecular biology have shown that coding DNA is not the only source of Crick information (see also Stotz 2006a, 2006b).

The idea of a genetic code is used more loosely in two ways. The first is talk of DNA 'coding for' RNA. The relationship between DNA and RNA is not mediated by the genetic triplet code. The ways in which DNA nucleotides pair with RNA nucleotides is determined by physics, and the units of the genetic code – the codons – play no role in the process of transcription by which a DNA molecule gives rise to an RNA molecule. Each of the four DNA nucleotides has a combination of hydrogen bonds which allow it to bind to one of the four RNA nucleotides. Sequences of three DNA nucleotides are no more significant to the machinery of transcription than sequences of two, four, or one hundred nucleotides. DNA contains 'codons' only in the sense that some parts of the DNA sequence correspond to RNA sequences which contain codons. A DNA sequence which does not correspond to a sequence of coding RNA cannot be meaningfully said to consist of three-nucleotide codons because there is no fact of the matter about where one codon ends and the next begins. The objection to talking of 'coding' in this context is that it adds nothing to what we already know – mapping the chemical process onto a communication system produces no insights. In fact, as Arnon Levy has pointed out to us, it suggests something false – that codons are functional units in the transcription process.

The translation of RNA into protein is also a chemical process, of course, but it is mediated by transfer RNAs, small 'adaptor molecules' one end of which has a chemical affinity for a codon and the other end of which has an affinity for an amino acid. In principle, any codon could be assigned to any amino acid by a suitably constructed transfer RNA. The standard genetic code (see Table 3.1) and the minor variations on that code, such as the code used to translate the mitochondrial DNA, are the result of which transfer RNAs are actually present. So the genetic code is a product of evolution and it makes sense to ask for an evolutionary explanation of the particular codon assignments. It would make no sense to ask for an evolutionary explanation of why a particular DNA nucleotide corresponds to a particular RNA nucleotide! This feature of the genetic code is important for two reasons, both of which will be explored below. First, it increases the strength of the analogy between the genetic code and human communication systems: the relationship between a symbol and its significance in human communication is usually arbitrary. Second, it means that theories about the efficiency of different coding schemes can be meaningfully applied to the genetic code: many different coding schemes using different tRNAs are possible, and they can be compared to the existing scheme. Although the idea of looking for the most efficient coding scheme did not prove useful in deciphering the genetic code in the 1960s, it has value in other contexts, as we will see below.

The second way in which 'genetic coding' is more loosely used is in talk of coding for phenotypes beyond protein structure. Outside molecular biology itself almost all talk of genetic coding is of this looser variety. DNA sequences are said to code for bodily features, for diseases, and for behaviours. This 'coding' cannot refer to the genetic code as we have discussed it so far. The genetic code is a language with very limited expressive power. It contains expressions that specify which amino acid is to be added to a polypeptide chain, and which instruct the ribosome to commence or terminate the process of adding amino acids. It cannot literally encode instructions for the development of more complex phenotypes, any more than I can give instructions on how to build a house using only expressions for latitude and longitude.

The fallacy implicit in extending 'genetic coding' in this way is made clear by an analogy due to Peter Godfrey-Smith (2000a). When I issue an instruction, many causal consequences may follow, connected to the instruction by more or less long and convoluted processes. But only some of these consequences correspond to what I instructed. When President Nixon ordered his staff to break in to the Democratic Party offices in the Watergate centre, he only ordered them to commit a break-in, not to end his presidency, or to cause a change in American attitudes to government. His words may have *caused* those things, but they did not *order* them. Similarly, although substituting one nucleotide for another may *cause* a change in my behaviour, as long as we are speaking strictly about the genetic code the change in my behaviour is not *coded* in the RNA transcribed from that locus. There may be independent reasons for describing the relationship between a DNA sequence and a phenotypic outcome in terms of 'coding', instruction', or 'information'. But these forms of informational language cannot be justified merely by pointing to the existence of the triplet code relating RNA to protein.

The development of the analogy between protein synthesis and the translation of a coded message can be usefully understood using Hesse's (1966) concepts of positive, neutral, and negative analogies. The analogy was introduced on the basis of some powerful positive analogies between the source and the target. The 1960s saw the exploration of many neutral analogies, many of which turned out to be negative analogies, as Sarkar (1996) has stressed. Eventually, the phrase 'genetic code' acquired a new, technical meaning which is exhaustively defined by the set of codon assignments that can be found in any textbook (see Figure 3.2). In technical contexts 'genetic code' no longer functions as analogy but simply refers to those codon assignments and the mechanism – the tRNAs – that underlies them.

6.4 The genetic programme

Perhaps the most contested aspect of information talk in biology is the idea that development is the expression of a programme written in the genes. This idea was central to the philosophical vision of biology developed by Ernst Mayr, the leading advocate of the neo-Darwinian 'Modern Synthesis' between genetics and evolutionary theory. Mayr used the idea of the genetic programme to explain the separate and complementary roles of the new, molecular biology and older biological disciplines such as his own home discipline of zoology (Beatty 1990). Molecular biology, along with older 'functional' fields such as embryology and physiology, studies the genetic programme as it exists today. But other fields within biology, such as systematics, population genetics, and much of traditional zoology and botany, study the creation of the genetic programme:

> We can use the language of information theory to attempt still another characterization of these two fields of biology. The functional biologist deals with all aspects of the decoding of the programmed information contained in the DNA code of the fertilized zygote. The evolutionary biologist, on the other hand, is interested in the history of these codes of information and in the laws that control the changes of these codes from generation to generation. (Mayr 1961, 1502)

The idea that development is the expression of a genetic programme has been understood in many ways, including that favoured by Rosenberg, in which the role of DNA in development is strongly analogous to the way in which a human designer might write a program for a self-replicating machine. At the other end of the spectrum, the genetic programme can be compared to the idea that the planets 'compute' their courses around the sun, integrating the gravitational forces which act on them so as to calculate their trajectory at the next instant. This is a picturesque way to talk, and might be a useful way to introduce the subject to students, but it is not a scientific insight into the nature of motion (Nijhout 1990).

The research tradition with the most evident bearing on how to understand talk of a genetic programme for development is the international collaborative effort to understand the development of multi-cellular organisms through an exhaustive analysis of the tiny nematode worm *Caenorhabditis elegans*. During this highly successful research programme, which began in 1974 and reached its apogee with the publication of the *C. elegans* genome in 1998, attitudes to the 'genetic programme' changed over time. At the beginning of the research programme it was an analogy which was expected to develop into a model of worm development that would reveal detailed resemblances to computing technologies. Later it was regarded as more of a metaphor for pedagogy or popularisation, and concerns arose about whether it was an apt metaphor for the actual models. Among the perceived disanalogies between the development of the worm and the execution of a computer program were the parallel rather than sequential nature of development, and the absence of a small set of repeated 'subroutines' out of which the 'program' was compiled (Chadarevian 1998; Schaffner 1998). However, as we will see below, these features may be present at a lower level of analysis, and provide grounds for a model which treats some aspects of development as a program.

Many students of behavioural development were immediately sceptical of the genetic programme idea. Daniel S. Lehrman, a frequent sparring partner of Mayr, wrote that:

> although the idea that behavior patterns are 'blueprinted' or 'encoded' in the genome is a perfectly appropriate and instructive way of talking about certain problems of genetics and evolution, it does not in any way deal with the kinds of questions about behavioral development to which it is so often applied. (Lehrman 1970, 35)

According to Lehrman, evolutionary design is not found only in the sequence of DNA bases. Evolutionary design assumes the presence of specific aspects of the environment, and in many organisms those aspects of the environment are actively provided by the previous generation as part of their 'developmental niche' (see Chapter 5).

Mayr's favourite illustration of a trait which is genetically programmed and thus directly explained by evolution was the ability of the North American Brown-headed Cowbird, a nest parasite like the European cuckoo, to recognise its own species after being raised by the host species (Mayr 1961, 1502). Mayr thought it evident that information about what cowbirds look like and how they sing must be programmed in the cowbird genome: 'The Gestalt of his own species is firmly imbedded [sic] in the genetic program with which the cowbird is endowed from the very beginning. It is – at least in respect to species recognition – a completely closed program' (Mayr 1964, 940). Ironically, as we saw in Chapter 5, the cowbird has now become an icon for critics of the genetic programme concept. Research has revealed that neither the ability of cowbirds to reliably identify the appearance of other cowbirds nor their ability to prefer or to produce species-typical song, depends on the kind of rigid, endogenous process Mayr envisaged. Cowbird ecology, flock structure, cultural transmission, and various forms of 'learning' unite to produce these phenotypes (West and King 2008; see 5.6).

The process by which cowbirds develop is a miracle of evolutionary design, and genes are involved at every stage. The interaction between the cowbird and its developmental niche regulates genome expression, using the kinds of mechanisms discussed in Chapter 4 and Chapter 5, and thus builds the neural and other structures which support further interaction with the niche. For this cascade of events to occur, the cowbird genome must have just the right resources to respond to the factors in the niche. But the claim that these phenotypes are 'programmed in the genes' seems to be no more than a picturesque way to say that the process of cowbird development is the result of natural selection. This is just what Lehrman feared – the idea of genetic information acting as a Trojan horse which disguises an evolutionary explanation as a developmental explanation, and obscures the need for a real, mechanistic explanation of development.

6.5 Genetic information and information theory

Concepts of information can be divided into two rough categories – *causal* information concepts and *semantic* information concepts (Sterelny and Griffiths 1999). Causal conceptions of information are inspired by the mathematical theory of communication, often called simply 'information theory' (Shannon and Weaver 1949). Information theory is only concerned with the quantity of information in a physical system. The quantity of information in a system can be understood roughly as how much disorder it contains. Measuring the quantity of information is similar to measuring the entropy of a physical system in thermodynamics. Information theory says nothing about the content, or meaning, of the information. Information flows over a channel connecting two systems: a sender and a receiver. There is a channel between two systems when the state of one is systematically related to the other, so that the state of the receiver is correlated with the state of the sender. A signal carries information about a source when the state of the signal allows the receiver to reduce his or her uncertainty about the state of the source. The mathematical theory of communication provided a quantitative measure of how much that uncertainty is reduced by a given signal. It was designed to allow, for example, calculation of the most efficient coding scheme with which to transmit a given message via a channel with given characteristics, or of the maximum amount of information that can flow through a given channel given an optimal coding scheme. Although the formal theory does not deal with the meaning of signals, there is a way of thinking about meaning which is broadly in the spirit of information theory. The philosopher H. P. Grice called this 'natural meaning' (Grice 1957). The natural meaning, or causal information content, of a state of the receiver is a state of the sender with which it is correlated. Thus, smoke carries information about fire and disease phenotypes carry information about disease genes.

The obvious way to defend information talk in biology is to argue that it is just like the many other scientifically respectable uses of causal information. This has been the traditional strategy. One of the authors was present at an address to the Fifth International Congress of Evolutionary and Systematic Biology in 1996 when the influential evolutionary theorist John Maynard Smith made use of this conventional defence. Talk about genetic information, he said, was to be interpreted 'more or less in the spirit of information theory' (author's notes).

The disadvantage of this defence is that it implies parity between genetic and non-genetic causes in development. As Maynard Smith later wrote,

> [w]ith this definition, there is no difficulty in saying that a gene carries information about adult form; an individual with the gene for achondroplasia will have short arms and legs. But we can equally well say that a baby's environment carries information about growth; if it is malnourished, it will be underweight. (Maynard Smith 2000, 189)

Causal information is created by the systematic dependence of a receiver on a source. The states of affairs that create this dependence are referred to as 'channel conditions'. In the case of development, the genes can be a source, the life cycle of the organism can be a receiver, and the channel conditions can be all the other resources needed for the life cycle to unfold. But it is a fundamental feature of information theory that the role of source and channel condition can be reversed. The traditional television 'test-card' held the actual transmission constant so that the television engineer could read off the state of what were previously channel conditions, such as the antennae on the roof. The source/channel distinction is imposed on a natural causal system by the observer. A source is simply one channel condition whose current state the signal is being used to investigate. If all other resources are held constant, a life cycle can give us information about the genes, but if the genes are held constant, a life cycle can give us information about whichever other resource we decide to let vary. So far as causal information goes, every resource whose state affects development is a source of developmental information (Johnston 1987; Gray 1992; Griffiths and Gray 1994; Oyama 2000a).

This symmetry is an instance of the 'parity thesis', an idea originally due to Susan Oyama (1985, 2000b, 2–3) and elaborated by Paul Griffiths and Russell Gray (1994; see also 2005). The parity thesis asserts that the roles of causal factors in development do not fall neatly into two kinds, one role exclusively played by DNA and RNA sequences, and the other role exclusively played by elements other than nucleic acids. The roles played by DNA and RNA sequences in development are sometimes filled by other developmental resources. For example, Chapter 4 described how non-genetic factors can provide informational specificity, or Crick information as we called it, of just the kind that figured in Crick's sequence hypothesis. Conversely, some DNA sequences play roles more usually associated with a non-genetic factor. For example, chromatin insulator regions of DNA play a role in the facultative modification of chromatin to regulate gene expression. Their role is more like the role of the cellular epigenetic mechanisms such as DNA methylation discussed in Chapter 5 than the role played by coding regions of the DNA.

The fact that causal information conforms to a 'parity thesis' has been accepted by most participants in the debate over how to understand genetic information (Godfrey-Smith 1999; Sterelny and Griffiths 1999; Maynard Smith 2000; Shea 2007b; Bergstrom and Rosvall 2009).

6.6 The semantic gene

For Maynard Smith the parity of genetic and other classes of developmental factors with respect to causal information was good reason to abandon the idea that genetic information is causal information. The value of the idea of genetic information, he thought, was to explain why genes are different from other developmental causes:

> [I]nformational language has been used to characterize genetic as opposed to environmental causes. I want now to try to justify this usage.
>
> I will argue that the distinction can be justified only if the concept of information is used in biology only for causes that have the property of intentionality [. . .] A DNA molecule has a particular sequence because it

specifies a particular protein, but a cloud is not black because it predicts rain. This element of intentionality comes from natural selection. (Maynard Smith 2000, 189–90)

We said above that concepts of information can be divided into *causal* information concepts and *semantic* information concepts. Human thought and language are supposed to contain information in the semantic or 'intentional' sense. An intentional state carries the same information irrespective of whether it stands in a correlative or causal relationship to its content, or 'intentional object'. The intentional object can be something which does not exist: I can have thoughts about mermaids. Philosophers distinguish between the causal object of a thought (e.g. the sea-cow which swam past a drunken sailor), and its intentional object (the mermaid which the sailor saw). It is because intentional states have intentional objects as well as causal objects that they can be false – they can contain information about states of affairs that do not obtain. Another way to look at the distinctive nature of semantic information is that it is *normative*: indicative statements have truth-conditions, imperatives have compliance conditions, and so forth. Each statement has a corresponding state of affairs which may or may not exist, but which is nonetheless the normative condition against which the success of the statement is judged. Because the medieval legend about a woman disguised as a man becoming pope was false, the statement 'Pope Joan reigned in the twelfth century' is a failure. It does not succeed in doing what it is meant to do.

Intentionality is often said to be the feature which marks out mental states as distinctive from the rest of the natural world, an idea normally attributed to the nineteenth-century philosopher and psychologist Franz Brentano (1874). The question of how physical systems such as the brain can exhibit intentionality is one of the most vexed issues in contemporary philosophy. Naturalistically inclined philosophers would like to show that this puzzling phenomenon can be reductively explained in terms of less problematic features of the world. The obvious route would be to reduce semantic information to causal information, but it is widely agreed that attempts to do this have been unsuccessful. A signal cannot both correlate with a source and not correlate with it, nor can a signal correlate with a source that does not exist, so it is difficult to reproduce the phenomenon of misrepresentation using a causal notion of information (Godfrey-Smith 1989). The most promising attempts to give a naturalistic account of intentional information are the so-called 'teleosemantic' theories to be discussed below, according to which a representational state carries whatever semantic information evolution designed it to carry.

Genetic information is often described as if it made sense to speak of a phenotype misinterpreting the message in the genes. This strongly suggests that genetic information is being conceived as having intentionality, as Maynard Smith (2000) suggested in the quotation above. For example, when organisms develop in different ways depending on circumstances, this is often described as the result of a disjunctive genetic programme which responds to an environmental trigger by accessing one alternative branch of the programme. This is a familiar picture from nativist theories of the acquisition of language (Chomsky 1988). The genetic instruction takes the form 'Develop like *this* under these circumstances, like *that* under these other circumstances'. But no one says that the human genetic programme encodes the instruction 'When exposed to the drug thalidomide grow only rudimentary limbs'. If the information in the genetic programme were causal information, then this would be one disjunct of that genetic programme. When the relevant channel is contaminated by thalidomide, the human genome and its regulatory apparatus send this causal information. The fact that the notion of a disjunctive programme is not applied to outcomes that are thought to be pathological, accidental, or otherwise 'unintended' suggests that the information in the programme is being conceived as intentional information.

The suggestion that genetic information is intentional information raises the obvious and pressing problem of how a sequence of DNA could possess intentionality. No one wants to ascribe mental states to DNA sequences, and very few biologists would accept that DNA sequences derive their intentionality from God in the way that this sentence derives its intentionality from us, the authors. The only obvious solution is that embraced by Maynard Smith, which is to derive the intentional properties of DNA sequences from their design by natural selection. This approach to semantic information is known as 'teleosemantics' because it seeks to derive semantic properties from evolutionary design, or teleology. It suggests that an indicative representation, for example, is true just in case it is produced in accordance with the evolutionary design of the system that produces it. The truth condition of that representation is the state of affairs in the context of which the representation-producing system would produce such a representation if it were functioning correctly. The truth condition is how the world would have to be if the representation were to achieve its purpose: my belief that porcini mushrooms are edible will only fulfil its biological purpose if they are, in fact, edible. Parallel accounts can be given of imperatives, interrogatives, and so on. The first proponents of teleosemantics were philosophers hoping to explain the intentionality of human thought and language (Millikan 1984; Papineau 1987). The complex literature on that topic is not something we need to engage with here, and our attention will be confined to the relatively modest attempt to explain why it is possible to ascribe limited forms of intentionality to DNA sequences using teleosemantics.

Most advocates of teleosemantic genetic information concede that some other developmental causes, as well as genes, carry teleosemantic information. This is an immediate consequence of the existence of evolved mechanisms of non-genetic inheritance (see Chapter 5). If genes carry teleosemantic information because they have been designed to influence the development of offspring, then anything else designed to do that must also carry teleosemantic information (Sterelny et al. 1996; see also Griffiths and Gray 1997; Shea 2007b). This is to concede a version of the parity thesis. The distinction between genetic, informational causes and environmental, merely physical causes is replaced by a distinction between causes which are designed to carry developmental information and causes which are not designed to do this. The major histocompatibility complex genes and the antibodies in mothers' milk both carry teleosemantic information, because they are both designed to influence the development of the immune system. But the sunlight that prevents a child developing rickets and the gene for achondroplasia do not carry teleosemantic information because they were not designed to produce, respectively, healthy bones or short stature.

The authors who advocate a teleosemantic approach to genetic information have tried to limit the significance of this parity thesis by stressing that not all environmental factors which influence development carry teleosemantic information. Only environmental influences produced by a mechanism which has been designed by evolution to influence development are on a par with genes. Work on epigenetic inheritance in the past two decades, documented in Chapter 5, has made increasingly significant the concession that some non-genetic resources carry teleosemantic information. It was not necessary to understand in detail the molecular basis of development to appreciate the importance of the genetic heredity system. Similarly, the importance of exogenetic heredity systems 'parallel' to genetic heredity (see Chapter 5) has become evident even before the molecular details are fully elucidated. In 5.4 we introduced the idea of parental effects: correlations between parent phenotype and offspring phenotype which are independent of correlations between, on the one hand, parent genotype and offspring genotype, and, on the other, between parental environment and offspring environment. That is to say, they are the signature of some mechanism other than genetic inheritance by which parents systematically influence the phenotype of their offspring. The mechanisms underlying parental effects range from varying the mass of the egg yolk, to providing sophisticated forms of childcare. We described some examples in Chapter 5 and will describe another in a moment. What matters is the broad finding that parental effects exist in a wide range of plants and animals, that they have a significant effect on fitness, and that many of them are likely to be adaptations (for reviews, see Mousseau and Fox 1998; Uller 2008). It follows from this that there is a great deal of non-genetic teleosemantic information.

6.7 Teleosemantic transmission information

In the past decade a new and powerful defence of the scientific value of treating genetic and other causes in development as signals carrying information has emerged. This is sometimes referred to as the 'transmission sense of information' (Bergstrom and Rosvall 2009). The proposal is that genetic information finds its primary use in understanding the evolution of heredity systems. The genetic code is an adaptation for transmitting biological specificity from one generation to the next, and the efficiency of this adaptation can be seen by analysing it as an information channel using the mathematical theory of communication. In Bergstrom and Rosvall's own work, this idea is actually in the spirit of the 'syntactic' approaches to information discussed in 6.10. In this section, however, we will concentrate on the use of this idea to defend a semantic view of genetic information.

In an early presentation of the idea Eva Jablonka (2002) argued that a developmental cause carries information if there is a 'receiver' which systematically alters its own state in response to inputs of this kind, and this differential response by the receiver is 'functional'. If 'functional' is taken to imply that the receiver has been designed by natural selection to respond to the signal in this way, then this is a form of teleosemantics. Like Bergstrom and Rosvall (2009), Jablonka argued that the value of treating heredity as a flow of information is that we can compare the properties of different heredity

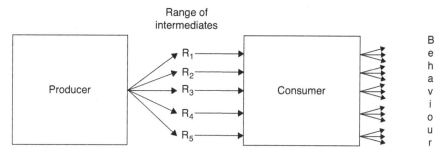

Figure 6.1 Infotel semantics. (Printed with permission from Oxford University Press: Nicholas Shea, in press. Inherited Representations are Read in Development. *British Journal for the Philosophy of Science*.)

systems and assess the selective advantages of one form of heredity compared to another.

In this section, however, we will concentrate on the prominent 'infotel' theory of Nicholas Shea (2007a, 2007b, 2011a, 2011b, in press). The infotel theory starts from the idea that genetic transmission is a way for organisms to send signals to their offspring, and seeks to define the semantic content of those signals. It combines teleosemantics with a requirement that developmental causes carry correlational information about the developmental environment. We will refer to information in Shea's sense as 'teleosemantic transmission information' in recognition of the fact that it represents a distinctive version of the teleosemantic approach, and also differs from Bergstrom and Rosvall's (2009) 'transmission sense of information'.

Shea defines the teleosemantic content of signals in the context of a 'representing system' like that depicted in Figure 6.1. It consists of a 'producer' which can produce a range of 'intermediates' R_i and a 'consumer' which can produce a range of behaviours. The behaviour of the consumer depends on which intermediate state is produced. According to Shea, if there is a system with this structure, then the conditions for an intermediate state R to have semantic content are as follows:

Tokens of type R have indicative content C if:

- Rs carry the *correlational information* that condition C obtains;
- an evolutionary explanation of the current existence of the representing system adverts to Rs having carried the correlational information that condition C obtains; and

- C is the *evolutionary success condition*, specific to Rs, of the output of the consumer system prompted by Rs. That is, C is the background environmental condition that made producing that output adaptive for a consumer in the past.

It is easy to see how these conditions are satisfied by an environmental input to an evolved mechanism of phenotypic plasticity. For example, some water fleas of the genus *Daphnia* can develop down two alternative developmental pathways. One of these involves the production of a defensive morphology which makes the flea more resistant to predation. This pathway is triggered by the presence of chemical traces of predators (kairomones) during development (Gilbert and Epel 2009, 27–8). This can be described as one of Shea's representing systems. The mechanisms by which embryos detect kairomones are the producer. The mechanisms that allow the embryo to develop down either developmental pathway are the receiver. The molecular signal produced by the system in the flea which detects kairomones (the producer) meets Shea's three conditions:

- It carries the *correlational information* that predators are present.
- The representing system evolved because kairomones are correlated with predation.
- The presence of predators is the relevant *evolutionary success condition*. The consumer system is designed to put the embryo down one developmental pathway when predators are present and another when they are not.

So according to the infotel theory, that molecular signal carries semantic information along the lines of 'predators present, grow defences'. The defensive morphology can also be produced as a parental effect, in which offspring of fleas with the defensive morphology develop that morphology without themselves being exposed to kairomones. In this case, the molecular signal by which the mother induces this developmental pathway in offspring will carry the semantic information 'predators present, grow defences'.

It is more complicated to apply Shea's model to genetic heredity (Figure 6.2). The first thing to grasp is that the representation system is partly at the population level and partly at the individual level. The producer system is the selective history of a *lineage* of organisms, so it exists at the level of the whole population. But the consumer system is an individual developmental process – there is a separate consumer system for each individual organism.

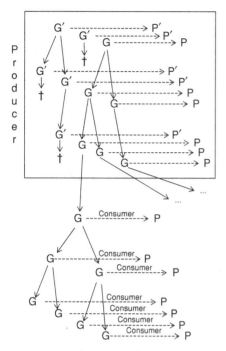

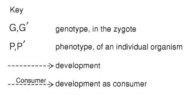

Key

G,G'   genotype, in the zygote

P,P'   phenotype, of an individual organism

----------> development

--Consumer--> development as consumer

Production of representation G takes place over many generations and involves selection. During this time, development (of phenotype P in response to genotype G) is part of the process of representation production. At the outset, G arises by a random process (e.g. mutation) and leads to, and correlates with, phenotype P; at the end of representation production G correlates both with P and with some environmental factor(s) E in virtue of which G was selected (E not shown).

Consumption of representation G takes place in each subsequent generation, in every individual carrying G in the zygote and developing P as a result.

Content of G is: *E is the case; develop P.*

Figure 6.2 Infotel semantics applied to genetic heredity. (Printed with permission from Oxford University Press: Shea, Nicholas, in press. Inherited Representations are Read in Development. *British Journal for the Philosophy of Science.*)

The intermediate states (representations) which the producer sends to the consumers are individual DNA sequences. In Figure 6.2 a particular genetic variant has gone to fixation, but this does not seem to be an essential feature of the model.

Natural selection leads to one DNA sequence (G') being eliminated (†) from the population while another (G) remains. Shea interprets this as the producer system producing one representation rather than another. The development of each individual is affected by receiving G instead of G'. Shea interprets this as the consumer producing a specific behaviour in response to a specific representation. According to Shea, G has semantic content because it meets his three conditions. First, it carries correlational information about the selection pressures in past environments: which sequences make it through depends on selection. Second, the entire representing system evolved because DNA sequences are correlated with selection pressures in past environments. This is a broad claim about the evolution of nucleic acid-based heredity systems which Shea defends at length. Third, he claims that the selection pressures in past environments, whatever they were, are the evolutionary success condition for G. He argues that the evolutionary explanation of the fact that developmental systems respond differentially to different DNA sequences is because this allows them to match phenotypes to the likely selection pressures in their environment.

We have noted that the 'representation system' flips between the population and individual level. This feature is shared by Maynard Smith's (2000) account, in which genes carry teleosemantic information because the process of natural selection is analogous to the design of computer code by a 'genetic algorithm'. The information is 'programmed' at the population level but 'read' at the individual level. There is an air of artificiality about this application of infotel semantics. The 'representation producer' is a population under selection, and the representation consumer is 'development', not a specific mechanism but whatever processes interact to produce covariance between the DNA sequence and the phenotype (Godfrey-Smith 2011, 180). But the force of both these criticisms depends on interpreting the model as a straightforward description, so that the 'producer system' and 'consumer system' need to be constituents of the organism on a par with the lymphatic system. However, as Tudor Baetu (2012) has pointed out, when scientists propose a model like this they typically use the language of analogy – the biological process is 'like' a signalling system, or it can be represented as, or modelled as, a signalling system. If the infotel theory is seen as a modelling exercise the artificiality of the genetic producer and consumer systems will not be an objection if modelling things in this way produces interesting results.

Arnon Levy has been more critical of the infotel theory, arguing that informational language is ubiquitous in many areas of biology, and not just in the contexts licensed by the infotel framework. He argues that informational language in biology should be interpreted in a 'fictionalist' manner (Levy 2011). Scientists notice that a biological process is analogous to communication, and use this analogy to construct a model in which one cell is communicating with another, or a transcription factor is signalling to a genetic locus telling it to initiate transcription. This is a fiction, because the intentional, communicative acts ascribed to cells and molecules are not acts that cells or molecules can commit, and the scientist knows that. But it is also an accurate representation of the biological process because the analogy draws attention to objective features of the biological process:

If one treats the process by which the pancreas controls glucose metabolism in informational terms, one is then obliged to designate the pancreas as a sender, insulin as the signal and so on. In other words, one may choose to view the process as an instance of intra-bodily communication, but it is not up to one what (informational) description the various elements should then receive. This is because the informational language serves as a way of pointing to the real (literally true) causal roles of those elements [...] (Levy 2011, 652)

But it is also open to the fictionalist to accept the infotel theory as one more fictional use of informational language. It models heredity on an analogy with communication. Shea is trying to show that genes and other developmental factors really have semantic properties, something Levy calls 'taking information too seriously' (Levy 2011, 652), but Levy can acknowledge the value of the infotel theory work in the same way he does that of other theorists who in his view 'take information too seriously'. The infotel theory draws attention to some objective features of the relationship between what genes do in development and the selection pressures that led to their evolution, using a fiction in which the developing organism is getting genetic messages from a population of its ancestors.

However, we do not think fictionalism is the best way to understand the transmission sense of information. The earlier authors who discussed the role of metaphor in science, such as Hesse, knew that it is possible to use metaphor and to engage in analogical reasoning, without adopting the pretence that the system you are reasoning about *is* the thing with which you are analogising it. Only some of the properties of the analogue are projected onto the target system. As well as positive analogies between the two systems (the properties we believe they share), there are negative analogies (the properties we believe differ), and neutral analogies (where we don't know if a property is shared by the target system). Further investigation of neutral analogies is a tool for scientific discovery (Hesse 1966).

Biologists who describe molecules as 'signals' do not need to pretend that these molecular signals have all the properties of signals in human communication systems. They can recognise negative analogies as well as positive analogies. Moreover, drawing an analogy is only the beginning of building a model. A mature model may use the vocabulary drawn from the analogy in new technical senses which include only the properties actually present in the target system. As a result of this kind of process, some information talk in biology is as literal as any scientific representation can be. One example is the genetic code. This encodes what we have called Crick information, the causal determination of the specificity of a biomolecule, and it is a code because it encodes that information in the formal sense of Shannon's theory of communication (Bergstrom and Rosvall 2009). If we calculate the redundancy in a coding sequence of DNA we are not creating a fiction in which a genetic agent beats about the bush when placing an order with the ribosome for a protein. The redundancy is a straightforward measure of whether that protein could have been specified in an alternative code using less DNA.

Shea's infotel theory could potentially be interpreted in this way, so that his 'inherited information' has only the properties formally justified by the aspects of the target system that are captured by his model. This will vindicate Shea's realism about inherited information, but at the cost that inherited information can only be used to answer 'ultimate' or evolutionary questions about inherited phenotypes, and not 'proximate' questions, such as how genes influence those phenotypes.

6.8 The limitations of teleosemantic information

Teleosemantics derives meaning from the evolutionary purpose of states that are said to carry teleosemantic information. So it is not surprising that teleosemantic information primarily answers an evolutionary question. That question is why organisms have phenotypes which are well designed for environments of which they had no experience during their development. The loose, metaphorical version of the explanation is that their ancestors experienced those environments, learnt about them, and passed on their knowledge. Something like this is implicit in Mayr's concept of the genetic programme, but the analogy was developed in more detail by the ethologist Konrad Lorenz (1965). The fact that an organism has an adaptive phenotype, Lorenz argued, shows that it has information about the demands the environment will place upon it. Lorenz distinguished two ways in which an organism could obtain such information: 'ontogenetic learning' and 'phylogenetic learning'. Ontogenetic learning includes learning in its traditional sense as well as other forms of developmental plasticity: the formation of calluses, for example, is 'learning' where the skin needs protection. Phylogenetic learning is the process of natural selection, which Lorenz regarded as a form of trial-and-error learning by a lineage of organisms. Lorenz argued that since organisms reveal this 'phyloge-netic information' in their adaptations, that information must be transmitted to them from their ancestors, presumably in their genes. Shea's infotel theory is very much in the same spirit.

A few years later the developmental psychobiologist Daniel S. Lehrman responded that 'although the idea that behavior patterns are "blueprinted" or "encoded" in the genome is a perfectly appropriate and instructive way of talking about certain problems of genetics and evolution, it does not in any way deal with the kinds of questions about behavioral development to which it is so often applied.' (Lehrman 1970, 35). The question he had in mind was that of his own research: what are the mechanistic processes by which a developmental system consisting of the fertilised egg and its developmen-tal niche constructs phenotypes? At first sight, there is something puzzling about Lehrman's statement. He accepts Lorenz's explanation of why offspring reproduce the phenotypes that allowed their ancestors to survive in terms of the transmission of genetic information. But he denies that this information explains the development of the phenotype in the offspring.

However, if we ask what Lorenz (1965) means by 'phylogenetic information' it becomes clear why it cannot explain development in a mechanistic sense. The phylogenetic information in a DNA sequence is the adaptive purpose which caused it to be selected: Lorenz is offering a version of teleosemantics. Teleosemantic information cannot play a role in a mechanistic explanation of how development unfolds because the teleosemantic content of DNA depends on its evolutionary history. A physically identical DNA sequence that arises by mutation in a new individual does not carry that information, because it does not have that history. But two otherwise similar organisms that have the same change in their DNA will be affected in exactly the same way. That the two sequences have different histories is simply irrelevant. The teleosemantic content of a DNA sequence makes no difference to what effect it has on development. Many authors have made this point (e.g. Shea 2007b, 318–19): explaining development with teleosemantic information is trying to answer a 'proximal biology' question with an 'ultimate biology' answer.

Nevertheless, the idea that teleosemantic information is somehow 'decoded' or 'read' by the developing organism remains attractive to many theorists. As we will see below, even Shea is tempted by it. We think this temptation is simply the result of equivocation between different senses of 'information', something that is only to be expected when the same analogy has been developed in several different directions. Shea's infotel semantics is, as he has noted, equally applicable to genetic and exogenetic inheritance mechanisms. With some exogenetic inheritance mechanisms there is less temptation to equivocate, and the error of using teleosemantic information in a mechanistic explanation is starkly obvious.

The North American seed beetle *Stator limbatus* follows alternative developmental pathways in response to the challenges posed by the seeds of different species. Eggs laid on seeds of the Catclaw Acacia (*Acacia greggii*) have very high rates of survival. Seeds of the Blue Palo Verde (*Cercidium floridum*) pose more of a challenge. In order to have a reasonable probability of survival when laid on palo verde seeds, offspring must grow faster and attain a larger final size than those developing on the acacia seeds. The choice of strategies is a 'parental effect' (see 5.4). Mothers lay fewer, larger eggs on the palo verde seeds than they do on the acacia seeds, and the egg mass causes the required form of development (Fox et al. 1995; Fox et al. 1997; we owe this example to Tobias Uller).

Phenomena like this can be modelled as the mother detecting something about the environment and signalling to the offspring. Having detected which kind of seed it is depositing eggs upon, the mother signals to the offspring to adopt one growth strategy rather than another. Using the infotel theory, we can say that the larger egg mass has the imperative content 'Grow fast and get large'. We can also use Shea's apparatus to assign to the larger egg mass the indicative content 'You are on *Cercidium floridum*'. In Shea's terms this is the 'evolutionary success condition' of the response of the 'consumer mechanism' to the egg mass.

However, this teleosemantic information does not translate into a mechanistic explanation of development. If we ask 'How does the egg mass produce faster growth and larger size?' and answer 'By transmitting to the mechanisms of development the instruction to grow fast and get large', this is no answer at all. Neither is 'By transmitting to the mechanisms of development the information that the egg has been laid on *Cercidium floridum*'.

But when the inter-generational signal is a DNA sequence, or even a methylation pattern on a DNA sequence, the vacuity of the explanation is less obvious. In such cases it is easy to think that the teleosemantic information also plays a mechanistic role in development when the information in that DNA sequence is transcribed. But this is to equivocate on 'information'. Whatever teleosemantic information a DNA coding sequence carries, it will still carry its usual payload of Crick information – the specification of sequence in the corresponding protein. But this is not the same information. First, the Crick information is not 'Grow fast and get large', but the specification of the order of amino acids in a polypeptide. Second, the Crick information in that sequence would be the same whatever its adaptive history, and whatever phenotypic effect was produced by the protein in another context.

In a recent paper, 'Inherited Representations are Read in Development', Shea (in press) appears to have found a way around these objections. Inherited information provides not only an evolutionary explanation of the trait, but also a mechanistic explanation of how it develops:

> We can distinguish two broad questions that can be asked about an individual
> episode of development: why did it arrive at a particular outcome; and how
> did the process unfold? This section focuses on the former, arguing that
> genetic representation explains some of the cases in which the outcome
> matches a feature of the organism's environment. We return in section 6 to
> questions about how the developmental process operates. (Shea, in press,
> Section 3)

But in fact this is a subtler conflation of two different senses in which we can 'explain development'. The examples Shea gives in which teleosemantic information 'explains development' are evolutionary explanations. They point to the adaptive advantages of certain developmental mechanisms and propose that this was an important factor in the historical evolution of such mechanisms. But an evolutionary explanation of a development mechanism is not the same thing as a mechanistic explanation of development. To see the difference it is useful to map the explanations into Tinbergen's 'four questions' framework (Tinbergen, 1963). This is a more detailed taxonomy of biological explanations than Mayr's (1961) proximate/ultimate distinction. A full biological understanding of a trait involves answering four questions:

1. Causation: what is the mechanism by which the trait produces its effects?
2. Survival value: how does the trait contribute to the organism's fitness? – 'how survival is promoted and whether it is promoted better by the observed process than by slightly different processes.' (Tinbergen 1963–, 418)
3. Ontogeny: how is the trait constructed in development?
4. Evolution: 'the elucidation of the course evolution must be assumed to have taken, and the unravelling of its dynamics.' (428)

Shea poses two questions of his own which he says can be addressed using teleosemantic transmission information. The first is why development has the specific outcome that it does. This is Tinbergen's fourth, evolutionary question. Shea answers this by pointing to adaptive advantages conferred by certain developmental mechanisms and proposing that these adaptive advantages were an important factor in the historical evolution of those mechanisms. Shea's second question is supposed to be 'How did the [*developmental*] process unfold?' (Shea, in press, Section 3). This sounds as if it will be Tinbergen's third, ontogeny question. But the question Shea actually answers is an evolutionary one: '[T]he informational perspective can help explain why the internal mechanisms of development – developmental programs, somatic cell inheritance, etc. – take the form that they do' (Section 6.2). So what Shea actually provides are not mechanistic explanations of how phenotypes are constructed by the regulated expression of the genome, but evolutionary explanations of why development uses a particular mechanism to produce that outcome. They are evolutionary explanations of developmental phenotypes.

Shea (in press) has identified two different questions, and one of them is a question about development. But they are not an evolutionary and a developmental question about the same phenotype. They are an evolutionary question about the original trait, and another evolutionary question about a different phenotype, a feature of how the first trait develops. The ways in which an organism develops – the fact that the early embryo forms a blastula and then folds in on itself to form a gastrula, or the fact that cell condensations form in the limb bud – are as much phenotypes as the relatively stable adult morphology which result from these processes, and the four questions apply to these developmental phenotypes as much as to any others.

6.9 Genomic programmes without semantic information

In an earlier paper, one of us identified yet another equivocation on different senses of 'information'. By juxtaposing a discussion of the actual genetic code with a discussion of gene control networks the impression is given that messages written in the genetic code are flowing through those networks (Griffiths 2001). Here are some sequential quotes from one section of the Maynard Smith paper discussed in 6.6:

There is, I think, no serious objection to speaking of a genetic code, or to asserting that the gene codes for the sequence of amino acids in a protein.

[...]

However, an organism is more than a bag of specific proteins. Development requires that different proteins be made at different times, in different places. A revolution is now taking place in our understanding of this process. The picture that is emerging is one of a complex hierarchy of genes regulating the activity of other genes. Today, the notion of genes sending signals to other genes is as central as the notion of a genetic code was forty years ago.

[...]

Informational terminology is invading developmental biology, as it earlier invaded molecular biology. In the next section I try to justify this usage. (Maynard Smith 2000, 187–9)

As we have seen, Maynard Smith goes on to construct a teleosemantic account of the developmental information of genes. Maynard Smith is quite correct to draw attention to the central importance of genomic regulatory networks in current efforts to understand development (Davidson 2001; Ptashne and Gann 2002). But gene regulatory networks do not explain development by explaining how messages written in the genetic code are decoded, nor how evolutionary meaning is deciphered by the mechanisms of development.

Genomic regulatory networks explain development in a straightforwardly mechanistic way, by explaining how genes are switched on and off in response to inputs from elsewhere in the genome and, as was explained in Chapter 5, in response to inputs from the developmental niche. The operation of these networks cannot be mechanistically explained by the evolutionary meaning of their inputs, any more than the mechanistic operation of provisioning in the seed beetle could be explained in this way in the previous section.

The gene regulatory network (GRN) depicted in Figure 6.3 is much smaller but no different in principle from the massive 'wiring diagrams' that represent real efforts to describe the interactions between genome sequences (Davidson 2001). This Finite State Linear Model (FSLM) consists of a set of parts (the three genes, their products, their binding sites, and the affinities of the products for the binding sites), a topology (a directed graph representing how the genes influence each other), and a control logic (in this model the Boolean functions AND, NOT, but in other models differential equations or a combination of the

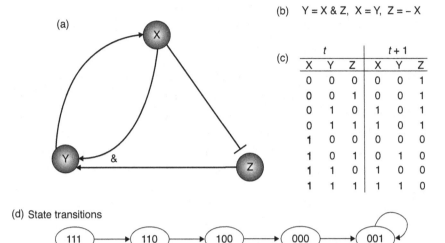

(a)

(b) Y = X & Z, X = Y, Z = – X

(c)

	t			t + 1	
X	Y	Z	X	Y	Z
0	0	0	0	0	1
0	0	1	0	0	1
0	1	0	1	0	1
0	1	1	1	0	1
1	0	0	0	0	0
1	0	1	0	1	0
1	1	0	1	0	0
1	1	1	1	1	0

(d) State transitions

111 → 110 → 100 → 000 → 001

011 → 101 ← 010

Figure 6.3 Example of a small Boolean network consisting of three genes: X, Y, and Z. Four different ways to represent the network are shown: (a) a graph, (b) Boolean rules for state transitions, (c) a complete table of all possible states before and after transition, (d) a graph representing the state transitions. (Reprinted with permission from The Royal Society: Thomas Schlitt and Brazma Alvis, 2006. Modelling in molecular biology: describing transcription regulatory networks at different scales. *Philosophical Transactions of the Royal Society* B 361: 483–94.)

two), and a dynamics (all the genes change state simultaneously in response to their previous inputs). The state-space of this model has eight states (see Figure 6.3c or d) and the dynamics of the model generate two attractors in this state space – 001 and a stable oscillation between 101 and 010 (Figure 6.3d).

This simple model is all we need to make the required points about information in GRNs. The model explains the behaviour of the network in the familiar mechanistic manner described in earlier chapters. The organisation of the parts explains why, from any given initial state, the network settles into a particular attractor state and remains there.

It is evident that the network is fundamentally cybernetic rather than semantic – if there are messages that flow through these networks they take the form 'switch on', switch off', 'upregulate', or 'bind' rather than 'predators present' or 'grow defences'. The operation of the mechanism is explained by two things, the specificity of the biomolecules involved and the fact that the interactions mediated by those specificities can be arranged into modules which implement logical functions. The best studied circuits of this kind are based on *cis*-acting elements:

> [C]is-regulatory modules that control developmental gene expression process the regulatory inputs provided by the transcription factors for which they contain specific target sites. A prominent class of cis-regulatory processing functions can be modeled as logic operations. Many of these are combinatorial because they are mediated by multiple sites, although others are unitary. (Istrail and Davidson 2005, 4954)

However, as we described in Chapter 4, the complete regulatory architecture of the genome extends to include *trans*-acting regulatory elements. It is the mechanisms of regulated recruitment and combinatorial control described in Chapter 4 which make up more complex genomic regulatory networks: regulatory elements in the genome recruit complexes of molecules which together determine the transcription and processing of gene products. GRNs also respond to external drivers in the developmental niche, as we described in Chapter 5. Because of the existence of exogenetic heredity, many of those drivers are the products of evolution and exist to provide the right developmental input at the right time. They are part of the evolved regulatory apparatus, although in current models they are typically treated as exogenous.

The difference between these genomic programmes and the genetic programme originally envisaged by Ernst Mayr has been emphasised by Tudor Baetu (2012). Baetu does not accept Rosenberg's grand vision of the genetic programme:

> the order in which the parts of the embryo are built is represented in the nucleic acid sequence [... and ...] what each component is made of and does is represented in the same sequence. (Rosenberg 2006, 95; cited in Baetu 2012, 659)

Baetu replaces this vision with a more prosaic picture:

> [...] genomic programs do not aim to show that DNA sequence X is 'the information source for' (or 'the cause of' in the case of reductionistic interpretations) phenotype Y, but rather to represent organizational features of the genome that enable it to contribute to a certain outcome (pattern of genome expression and, if known, associated trait, phenotype, or biological function/dysfunction). (Baetu 2012, 663)

Mayr's and Rosenberg's genetic programme, the blueprint for the phenotype, is not a mechanistic account of development, but a projection of evolutionary design. It is true that the system has been designed to produce evolved phenotypes, but, as we saw in 6.8, that does not translate into a mechanistic explanation of development. It merely poses the question of how development produces those phenotypes. Real GRNs are, as Baetu (2012) argues, mechanisms descriptions that contribute to answering that question.

6.9.1 Coding without semantics

In earlier work we have claimed that the genetic code is a form of causal information, a causal relationship between one physical state and another and no more intentional than Grice's 'natural meaning' (Griffiths 2001). We still endorse this conclusion, but with one major reservation. Bergstrom and Rosvall (2009) have argued convincingly that the genetic 'code' is a code in another sense, and that this is at least equally biologically significant. It is a solution to the adaptive problem of transmitting biological specificity from one generation to the next: a coding problem in the formal sense introduced by Claude Shannon (Shannon and Weaver 1949). The code – the mapping from nucleic acids to amino acids – has been shaped by natural selection for its efficiency in this role. Treating the coding relationship as something like Grice's 'natural meaning' – smoke means fire – pushes this feature into the background and underestimates the theoretical significance of interpreting these chemical relationships as a code. That significance, Bergstrom and Rosvall argue, lies not in a role for the code in describing or instructing phenotypes, where most philosophers have been looking, but its role in ensuring that the specificity required to construct those phenotypes is reliably transmitted from one generation to the next. Bergstrom and Rosvall emphasise that transmission information in this sense is not a semantic notion. The coding problem is posed – and solved – using Shannon's purely quantitative measure of information. What is being transmitted is specificity, but it is not necessary to know this in order to solve the coding problem.

6.10 Conclusion

In this chapter we have been sceptical of the search for meaning – semantics – in the genes. We have argued that the idea that genes are instructions for phenotypes, imposing form on matter in a semantic version of the preformation theory, results from confusing evolutionary, teleological explanations of development with mechanistic explanations. We have also continued to defend the 'parity thesis', one implication of which is that genetic and environmental causes of development are not distinguished by the fact that only genetic causes carry developmental information.

Once evolutionary explanations of development are distinguished from mechanistic explanations, the only grounds for denying that environmental factors in development 'carry information' are claims about the structure of the causal mechanisms of development. Waters' (2007) argument that coding sequences are the sole or main source of molecular specificity, with other causal factors in development at the cellular level having merely permissive causal roles, is the right kind of argument to refute the parity thesis (see 4.3). But while it is the right kind of argument, it is not a successful argument, as Stotz (2006a, 2006b) showed using the arguments for molecular epigenesis which we have reiterated in Chapter 4 and Chapter 5. Factors outside the DNA sequence co-specify the precise sequence of the RNAs and polypeptides that will be derived from the DNA.

We have reinforced the conclusion already urged by several philosophers that the only really substantial sense in which the genes carry information is their role in templating for gene products. This is Crick information – the causally specific determination of the order of elements in a gene product. Waters has correctly identified the key issue as the source of this specificity, but in the postgenomic era coding sequences have turned out to be sources of template potential that is flexibly used by the systems of which they are part and this larger system contributes to the determination of specificity through the activation, selection, and creation of coding sequences. These processes contribute additional Crick information for gene products which amplifies the Crick information in coding sequences (Chapter 4). We do not see why these conclusions should be thought to show that genes are not important, or to fly in the face of the obvious importance of nucleic acid-based heredity in living systems. Nothing we have argued disputes the fact that the evolution of nucleic acid-based heredity is the key evolutionary innovation that allows cells to transmit specificity, and thus the catalytic capacity of the cell (other philosophers who see this as a key point include Moss 2003 and Sarkar 2005). Moreover, while we have argued against an exclusive reading of the sequence hypothesis, the fact that coding sequences are rich sources of Crick information is at the heart of the alternative account we have given.

In the previous section of this chapter we described some recent work which identifies substantive scientific work done by informational models in biology. We regard this as substantial theoretical progress in the philosophy of biology. The secret to its success is abandoning the idea that the value of informational language in biology is to vindicate a semantic preformationism in which development is explained by a representation of, or instructions for, developmental outcomes. We have identified important and distinctive features of DNA, but ones that leave the parity between genetic and other causes in mechanistic accounts of development unscathed.

Further reading

The most substantive historical account of the informational turn in genetics and molecular biology is Lily Kay's *Who Wrote the Book of Life: A History of the Genetic Code* (2000), although her account is not uncontroversial. The best short introduction to the semantic view of theories and its application to biology remains Paul Thompson's *The Structure of Biological Theories* (1989). A good, short introduction to the recent literature on models and modelling is Roman Frigg and Stephan Hartmann's article in the *Stanford Encyclopedia of Philosophy* (Frigg and Hartmann 2009).

7   The behavioural gene

7.1 Behaviour genetics

In this chapter we introduce the traditional, quantitative approach to understanding the role of genes and environment in behaviour and behavioural difference, and contrast it to the causal analysis of the interaction of genes and environment recommended by critics of traditional behaviour genetics, especially Gilbert Gottlieb.

We show that there are two very different ways in which responsibility for a behavioural difference can be sheeted home to a genetic difference. These involve two identities of the gene: the familiar Mendelian allele and another identity which we call the 'abstract developmental gene'. Using these different identities of the gene, and the conceptions of gene action which accompany them, we can throw light on some of the disputes between behaviour geneticists and their critics.

In the final section of the chapter we suggest that the abstract developmental gene is now becoming concrete, as particular sequences of DNA. Behaviour genetics has also started to locate the basis of the statistical component of genetic variance in DNA sequences. The result is a convergence of the role of genes in both approaches. This explanatory role for genes resembles that of the earlier, abstract developmental gene. It is a contextual, systems-style form of genetic explanation and constitutes a corrective to the idea of biological explanation as the identification of specific actual difference makers advocated by Waters (2007) (see 4.3).

Traditional behavioural genetics was the application of genetic analysis and of quantitative genetic methods to behavioural phenotypes. Two of the most significant figures in the history of classical genetics, Theodosius Dobzhansky and Seymour Benzer, turned their attention to the genetics of behaviour later in their careers, continuing to work with *Drosophila* and seeking to understand the role of genes in behaviour via the genetic analysis of behavioural characters. In human behavioural genetics, however, quantitative genetic methods predominated. Quantitative genetics was developed by the founders of modern population genetics to integrate the Mendelian model of inheritance with the earlier, biometrical tradition in the study of natural selection. In Chapter 2 we encountered the biometrician Karl Pearson, who argued that the biometric theory of heredity was no more than the application of statistical methods to predict offspring phenotypes from data about parental phenotype. Rather than analysing phenotypes into a set of discrete characters which would yield Mendelian ratios, the biometricians simply measured phenotypic characters and the correlations between those characters in relatives. Quantitative genetics is an elaboration of this approach. Ronald A. Fisher's seminal paper 'The Correlation between Relatives on the Supposition of Mendelian Inheritance' (1918) derived predictions for the correlation between relatives from a Mendelian model and argued that they were in close agreement with the empirical data available at the time. The result of this work was to re-ground the biometric approach on Mendelian assumptions.

Fisher's 1918 paper also introduced the new statistical concept of *variance*. In contrast to earlier, correlational analyses of similarity, Fisher's attention to variance allowed for analyses of difference. The variance of a character measures how much a character differs in a population. Variance is calculated by measuring the difference between the value of the character in each individual and the population average value. The variance is the mean of the squares of those differences. Fisher used this measure to quantify genetic differences and environmental differences and establish how much each contributed to the phenotypic differences in a population. This focus on the 'relative contributions' of genetics and environment became the defining methodological feature of traditional behavioural genetics.

The technique Fisher (1918) invented, the analysis of variance (ANOVA), remains a standard method for investigating the relative contributions of genotypic and environmental variation to total phenotypic variation in a population. In the simplest case ANOVA partitions the total phenotypic variance for a trait (V_P) into a contribution attributable to genotypic variation (V_G) and a contribution attributable to environmental variation (V_E):

$$V_P = V_G + V_E$$

The equation assumes that the two sources of variation are additive, meaning that the phenotypic variance V_P can be obtained by simply adding V_G and V_E together. We will return to this below.

The simplest measure in behavioural genetics is broad heritability (H^2), which is the proportion of the phenotypic variance attributable to genetic variance:

$$H^2 = V_G/V_P$$

But V_G (and therefore H^2) confounds a number of factors. Broad heritability includes parental effects (Chapter 5). It includes the independent effect of each allele on the trait, but also the effect of one allele being dominant over another. In a simple case of dominance, the recessive allele in a heterozygote makes no contribution to determining the variance. But if we want to predict the outcome of breeding we need to remember that recessive allele. Broad heritability does not do this.

Dominance implies that the effects of alleles are not additive: how an allele affects the phenotype depends on which other allele it is paired with. Narrow heritability (h^2) is the proportion of total phenotypic variance accounted for by just the additive effects of alleles (V_A):

$$h^2 = V_A/V_P.$$

Narrow heritability can be used to predict the actual outcomes of breeding. It is at the heart of the practical application of quantitative genetics to selective breeding in plants and animals. Calculating narrow heritability requires correcting for dominance and for several other factors. For example, genes interact with one another in their effect on the phenotype, generating more non-additive genetic variance. By the time factors such as parental effects or non-random mating have been accounted for the equation is a great deal more complex. The heritability equation is a model of all the sources of variation in the population and the adequacy of the measure of h^2 depends on the adequacy of that model.

The most famous complication in measuring heritability is *gene–environment interaction*. The effect of a genetic difference often depends on the environment in which the genes are expressed, and the effect of an environmental difference often depends on the genotype of the organism. To take a simple but not unrealistic example, suppose that a gene produces an enzyme which only works below a certain temperature. If two individuals, one of whom has this

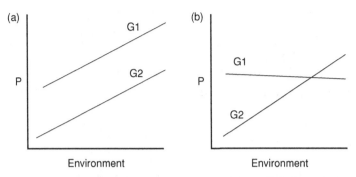

Figure 7.1 Hypothetical phenotypic curves: (a) parallel phenotypic curves with no gene–environment interaction; (b) non-parallel phenotypic curves with gene–environment interaction.

gene, are raised in environments below that critical temperature, then one will show the effects of the enzyme and the other will not. But if the two individuals are raised in environments above the critical temperature, then there will be no difference between them. So in this case, changing the environment eliminates a differences caused by a gene. In other cases, changing a gene can eliminate a difference caused by the environment. This means that the amount of genotypic variation is dependent on the distribution of genotypes across environments and vice versa. In statistical terms the two 'main effects' of genes and environment are joined by variance due to interaction or an 'interaction effect' ($V_{G \times E}$). The effects of V_G and V_E on V_P are no longer additive, so the interactions term must be added to the equation to get the total phenotypic variance:

$$V_P = V_G + V_E + V_{G \times E}$$

A good way to understand the difference between additive and non-additive interaction is to use norm of reaction figures, which graph the value of a phenotypic character against the value of an environmental factor for each genotype (Sarkar 1999). When V_G and V_E are additive, then the norms of reaction for two genotypes (G1 and G2) will be parallel across the whole range of environments (Figure 7.1a). In this figure there is no need to know what environment an organism lives in to predict how much phenotypic difference a genetic difference will make within the measured range of environments. However, when there is gene–environment interaction then the norms of reaction will not be parallel across the environments (Figure 7.1b). In this figure you need to know the environment to predict the effect of a genetic difference within the measured range of environments, and vice versa. You cannot identify the contributions of gene and environment independently and then add them up.

A point often made by critics of behaviour genetics is that predictions about what the norm of reaction will look like outside the measured environments are dangerous, since there may be no gene–environment interaction in measured environments but gene–environment interaction in nearby, unmeasured environments (or vice versa).

One practical implication of gene–environment interactions is that when they occur, heritability scores cannot be extrapolated from one population to another or from one environment to another. A trait can show high heritability in one population, but low heritability in another; likewise, it can show high heritability in one environment, but low heritability in another. Statistical interaction between genes and environment is well documented in behavioural genetic studies on animals (Fuller et al. 2005). The behaviour geneticist Douglas Wahlsten has been a long-term critic of human behavioural genetics for failing to find similar effects, which he attributes to methodological failings (Wahlsten 1990). This complex and long-running debate is dealt with in the works recommended at the end of the chapter.

Measures of heritability are not and do not pretend to be measures of whether the traits of individual organisms owe more to the genes or more to the environment: 'loose phrases about the "percentage of causation" which obscure the essential distinction between the individual and population should be avoided' (Fisher 1918, 399–400). To see why it is a mistake to take the heritability figure for a population as an indication of what genes and environment are doing in an individual consider the following: in a population of genetically identical individuals, all traits have a heritability of zero. But in each of those individuals the genes are doing the same things they would be doing if the individual lived in a more heterogeneous population. The zero heritability is irrelevant. Making people more genetically similar reduces heritability for the trivial reason that there are proportionally fewer genetic differences to correlate with any phenotypic differences. Conversely, making the environment more uniform *increases* heritability because there are proportionally fewer environmental differences to correlate with any phenotypic differences. For example, providing equal access to education for everyone in a community will increase the heritability of IQ in that population.

Heritability measures whether the actual differences seen in a population are due to actual differences in genes – it measures to what extent genetic differences are the causally specific actual difference makers (SADs) in Waters' sense (Waters 2007; see Chapter 4). But when we ask how genes are involved in producing a trait, we do not want to know whether they are specific *actual* difference makers; we want to know if they are specific *potential* difference makers. We want to know if the person would have been different in the specific respect that interests us had their genes been different (similar points are made in Kendler 2005 and Woodward 2010). As we will see, this distinction can be used to considerable effect to clarify some of the debates around behaviour genetics.

7.1.1 Assessing heritability

In agricultural genetics heritabilities can be calculated from controlled experimental data. In human behavioural genetics they must be inferred from observational data – that is to say, from the correlation between relatives. The differences in the genetic relationship between different classes of relatives, when compared to the differences in their phenotypes, allows the estimation of the genetic component of variance. Inferring causes from observational data is intrinsically difficult, and in meeting this challenge behaviour genetics has developed an immensely sophisticated tradition of the design and statistical analysis of observational studies. Much of the sophistication comes from looking for indirect ways to confirm the model underlying the estimation of heritability, and to estimate or control for the impact of the additional factors described above. Many of the most famous study designs involve comparisons between pairs of twins. One kind of twin study compares the correlation between monozygotic (MZ) twin pairs, who share all their genes (Mendelian alleles), and dizygotic (DZ) twin pairs who share only half their genes. If the MZ twins resemble one another more closely than the DZ twins, all other things being equal, this can be attributed to their greater proportion of shared genes. Broad heritability for a character can be calculated using the formula

$$H^2 = 2(r_{MZ} - r_{DZ})$$

where r is the coefficient of correlation between the twins of each kind. So H^2 is twice the difference between how similar MZ twins are to one another and how similar DZ twins are to one another. But all other things are not equal, as behaviour geneticists are well aware. Some characteristics in twins, such as IQ scores, show a strong maternal effect from the shared womb, and this is confounded in the measure of broad heritability (Devlin et al. 1997). The maternal effect is particularly relevant when explaining the similarity between twins adopted into different families. But straightforward MZ–DZ comparisons are also influenced by maternal effects. A large proportion of MZ twins share the same placenta, making their developmental environment more similar than that of DZ twins (Robert 2000). Serious estimates of heritability are far more complex than we can explain here, and disagreements about methods and about various parameters mean that estimates of the heritability of a specific character can often vary widely.

Nevertheless, traditional behaviour genetics is judged by most of its practitioners to have been a highly successful research tradition. In a well-known article, behaviour geneticist Erik Turkheimer summed up its findings as the three 'laws' of behaviour genetics:

> *First Law.* All human behavioral traits are heritable.
>
> *Second Law.* The effect of being raised in the same family is smaller than the effect of genes.
>
> *Third Law.* A substantial portion of the variation in complex human behavioral traits is not accounted for by the effects of genes or families. (Turkheimer 2000, 160)

The first point ('laws' was not meant seriously) is that for any behavioural character which varies in human populations, and for which a large enough dataset has been assembled, a portion of the phenotypic variance is statistically accounted for by genetic variance (i.e. all traits have a non-zero heritability). The second point is subtler. The similarity of the developmental environment measured using socio-economic status, divorce, and so forth does not explain as much of the variation between individuals as common sense would suggest it should. The failure to account for much variance using what behaviour geneticists call 'shared environment' may reflect the fact that we do not have a systematic theory of the human 'developmental niche' (5.6) to complement our knowledge of genetics. The proportion of Mendelian alleles shared between individuals is a matter that can be clearly established and it is a good measure of genetic variation. But the proportion of the environment shared between individuals is much less well defined.

Turkheimer's third point concerns what behaviour geneticists call 'non-shared environment'. This is a complex and contested concept (Schaffner 2006a, 30–2). Turkheimer distinguishes between the 'objective non-shared environment', the factors we can identify which differ between individuals, and the 'effective non-shared environment', which is simply a portion of the environmental variance. In earlier work Turkheimer had argued that the objective unshared environment accounts for very little variance. Here he suggests that the effective non-shared environment consists of a large number of relatively unsystematic influences on individuals, making the transition from identifying the statistical effect to identifying actual environmental causes difficult if not impossible. While this is by no means a universally accepted interpretation, it seems to be quite widely held. It is known as the 'gloomy prospect', from a remark by two behaviour geneticists who do not share this view:

> One gloomy prospect is that the salient environment might be unsystematic, idiosyncratic, or serendipitous events such as accidents, illnesses, or other traumas [...] Such capricious events, however, are likely to prove a dead end for research. More interesting heuristically are possible systematic sources of differences between families. (Plomin and Daniels 1987, 8; quoted in Turkheimer 2000)

In addition to this gloomy prospect about the environment, Turkheimer was prescient in suggesting another gloomy prospect for behavioural genetics. Writing just at the conclusion of the human genome project, he expressed scepticism that high heritabilities for complex human behavioural traits would turn out to be explained by a small number of genes each with a substantial effect on the behaviour. He raised the prospect that the genetic contribution to behaviour might be as complex as the environmental contribution.

7.1.2 Finding behavioural genes

Having established that a trait is heritable, it makes sense to try to locate the genes that vary in the population and whose variation explains the phenotypic variation. There are many methods to do this. The oldest is linkage analysis. In 2.2 we described how early classical geneticists developed 'linkage maps' of the probability that alleles will be inherited together. Linkage is roughly explained by physical proximity on a chromosome, since alleles which are closer together are less likely to be separated by crossing over in meiosis. The location on the chromosomes of the alleles for a trait can therefore be investigated by finding that it is linked to other traits for which the alleles are already known. Linkage studies require 'pedigrees' showing how the trait recurs in particular families. The recurrence of the trait in the pedigree can be correlated with the inheritance of known alleles, either by tracking the phenotypic marker of those alleles, or by genotyping the individuals to see directly which alleles they carry. In practice, of course, linkage analysis is a family of varied and complex study designs and of statistical methods for detecting linkage.

Association studies are another way to detect specific alleles responsible for the inheritance of a trait. The simplest version compares individuals who have the trait with those that do not (case-control method) and seeks to correlate this difference with presence of an allele, with a genetic marker, or with a haplotype (a distinctive portion of chromosome). With continuous traits like height or IQ score an association study assesses whether some portion of the variance is correlated with the genetic difference. Regions of the genome that account for some portion of the variance are known as quantitative trait loci (QTLs). Association studies raise obvious issues about confounding correlation and cause. For example, if a trait is distinctive of people with a particular geographic origin – blue eyes in Scandinavians – then it will correlate with other genetic variation that is distinctive of that population. Once again, actual work of this kind utilises varied and complex study designs and statistical methods which aim to overcome these issues.

In the postgenomic era it has become possible to conduct exhaustive searches for the genetic bases of phenotypic variance. The best-known method is a genome wide association study (GWAS), a variant on the traditional association study described above. The human genome contains around 10 million common variations at single nucleotide positions in the genome (single-nucleotide polymorphisms or SNPs). It does not matter whether these are of any functional significance – they simply act as a map of the genome. GWAS can be used to compare two populations, one with the trait and one without, to see which SNPs are statistically associated with the trait, and hence which regions of the genome are likely to contain the variation which accounts for phenotypic variance in the trait. GWAS can also be used to identify QTLs. GWAS is one of a growing family of methods. Given the sequence of the human genome, an increasingly detailed annotation of that sequence, and now databases containing many individual human genomes and documenting their differences, there are many ways to detect associations between genetic and phenotypic variation.

Traditional behaviour genetics was very successful in demonstrating a genetic basis for behavioural differences (hence Turkheimer's first law). But there has been little success in locating the molecular basis of that genetic component. It seems likely that Turkheimer was right, and the genetic basis of behaviour is as complex as its environmental basis. Techniques such as GWAS have been extensively used to investigate human behavioural traits, especially psychiatric disorders for obvious pragmatic reasons. The results typically reveal a large number of loci each of which accounts for a very small amount of variance, and which collectively account for much less of the variance than is thought to be genetic from traditional quantitative genetic analyses. This is known as the problem of 'missing heritability'. For example, human height has a heritability of around 80 per cent in European populations. In 2008 a series of GWAS studies revealed around forty SNPs associated with height (Visscher 2008). These accounted for around 5 per cent of the variance. In 2010 a study found some of the missing heritability for human height (Yang et al. 2010). Rather than try to find individual SNPs that accounted for a significant proportion of the variance they looked at the collective effect of many SNPs. Considering a little under 300,000 SNPs at once they were able to account for 45 per cent of the variance in human height. The editorial in that issue of *Nature Genetics* (2010) commented laconically that 'there are likely to be limits to the usefulness of the current strategy of accumulating common variants of small effect for risk prediction'. It is now reasonable to believe that the genetics of many complex behaviours, including the psychiatric disorders that have been so well studied, may follow this pattern (Manolio et al. 2009). The genetic bases of these traits, and of differences in these traits, are distributed across the genome in many loci, coding and non-coding, and in their interactions.

7.2 The dual identities of the behavioural gene

Behaviour genetics has always been controversial, because of concerns about its social and political implications. This has tended to distract attention away from more substantial scientific criticism. Some of the strongest criticism has come from developmental scientists, and especially developmental psychobiologists, a research tradition which emerged in the 1960s from work on behavioural development by comparative psychologists such as Daniel S. Lehrman, whose ideas were discussed in Chapter 6 (for an introduction to this kind of work and its relationship to behaviour genetics, see Hood et al. 2010). Developmental psychobiologists have been scathing about traditional, quantitative genetic approaches to behaviour, arguing that they do not yield genuine scientific understanding of the basis of behaviour. *They want to know how genes cause behaviour, not merely how much behaviour genes cause.* Traditional, quantitative genetic methods are fundamentally unsuited to the study of the causal role of genes in development because they analyse and explain phenomena at the level of the population and not the individual organism, and because they explain the differences between individuals, rather than how those individuals came to have the phenotypes that they do (Ford and Lerner 1992; Gottlieb 1995; Wahlsten and Gottlieb 1997; Meaney 2001a; Gottlieb 2003).

In this section we explore the very different conceptions of genes and of gene action that separate these disputants. This will exemplify how the many identities of the gene can cause problems when claims about genes move from one arena to another. We will also start to see that while behaviour geneticists are focused on the actual causes of variation, developmental scientists are interested in its potential causes.

Behaviour geneticists, and quantitative geneticists more generally, conceptualise genes in classical, Mendelian terms. While they may not conduct classical genetic analysis, their study designs and statistical models deal with the consequences of some number of Mendelian alleles, stretches of DNA whose inheritance explains the inheritance of phenotypic differences. In contrast, those who work in developmental science conceptualise genes as determinants of the value of a developmental variable in the context of a larger developmental system (we will refer to these constructs as 'abstract developmental genes'). In many instances we think that both sides would pick out the same specific DNA sequence elements if they had sufficient information about the molecular basis of a trait. But until very recently that information has not been available, and genes have existed for both communities only at the conceptual foundations of their actual methods. It is only in the last few years that behaviour geneticists have been able to identify the Mendelian alleles in their models with specific sequences of DNA, and developmental scientists have manipulated development by manipulating DNA. It is perhaps because of this lack of 'boundary objects' (Star and Griesemer 1998) which can move between the two intellectual contexts that there has been so much miscommunication.

Developmental science has sought to characterise what we called in 5.6 the 'developmental niche'. It investigates how normal development at each stage of the life cycle depends on the interaction of the organism with specific features of the environment. We described in 5.6 West and King's work on cowbirds, and Alberts' work on the rat. Another classic research exemplar is Celia Moore's work on penile reflexes in the rat (Moore 1984, 1992). Celia Moore showed that the spinal cord nuclei of male rats differ from those of female rats in ways that allow the male to use his penis during copulation. These neural differences result from differences in gene expression in the developing spinal cord of the rat pup, which in turn result from differences in the licking of the genital area by the mother, which in turn results from greater expression in male pups of a chemical that elicits maternal licking. In this research it was assumed that the environmental variables that are being manipulated by the experimenter exert their affects by modulating gene expression, but the research did not identify the genes involved. Recent work in developmental psychobiology has begun to link the parameters of developmental models to the expression of specific coding sequences in the genome (e.g. Meaney 2001b; Suomi 2004), but for most of the history of this research tradition genes have been highly theoretical entities. It has not been possible to manipulate specific genetic variables of the developmental system in the same way as specific environmental variables. Although developmental psychobiologists conceived of genes as mechanistic causes of development, the lack of direct access to these causes led them to appear in an extremely abstract form, simply as the determinants of the value of certain variables of a developmental system (Griffiths and Tabery 2008; Tabery and Griffiths 2010).

This 'abstract developmental' conception of the gene can be traced back to the 1930s when embryologists attempted to integrate genetics into their discipline. If it is assumed that the biochemical processes that construct phenotypes are the result of gene action, then some or all of the variables of a developmental model can be labelled as 'genes'. Julian Huxley speaks of 'rate genes' determining the value of variables in his models of relative growth

(Huxley 1972 [1932]). These hypothetical 'genes' have no empirical foundation besides the model itself and the general conviction that an organism's biochemistry is an expression of its genes. The same abstract conception appears in Waddington's classic representation of development as a complex system whose parameters are genetic loci and whose state space is a set of phenotypic states (Waddington 1940, 1957). In Figure 7.2 the state space is depicted as a surface, each point of which represents a phenotype (a). The genetic parameters are depicted as pegs that pull on the surface and thus determine its contours (so phenotypes are quite literally 'sheeted home' to genes)! Epistatic interactions between loci are represented by links between the cords by which those loci pull on the surface (b). The development of an organism over time is represented by the movement of a ball over the surface, which is dictated by gravity, so that the ball rolls downhill on a path dictated by the contours of the surface. The development of the organism is thus represented by its trajectory over the surface, through successive phenotypic states.

Mendelian alleles and abstract developmental genes are two legitimate ways to introduce DNA sequences into two very different research contexts. But the explanatory role which the 'gene' plays in those two contexts is very different. An abstract developmental gene can only explain a phenotype via the mediation of many other developmental variables. In contrast, the Mendelian allele for a phenotypic difference explains that difference without reference to other developmental variables. The abstract developmental gene has no identity apart from its role in a developmental model. In a model intended as an actual characterisation of a developmental process the introduction of specific 'genes' is justified by reference to the ability of the model as a whole to explain the effects of manipulations of its variables. So explaining the presence of a phenotype by reference to the presence of a gene means drawing attention to how that genetic variable interacts with the other variables. The same point applies to phenotypic differences, which are explained by reference to how a genetic difference ramifies through the system (Griffiths and Tabery 2008; Tabery and Griffiths 2010).

Explanations of phenotypes in terms of the presence of Mendelian alleles do not share these features. The presence of an allele (or of different alleles) explains the presence of the associated phenotype (or a phenotypic difference) because of a statistical association between alleles and phenotypes in a pedigree or a population. The epistemological value of this relationship derives precisely from the fact that it is robust across the actual distributions

(a)

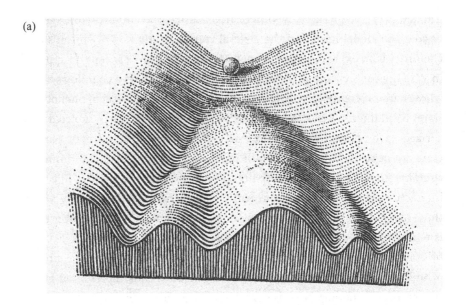

(b)

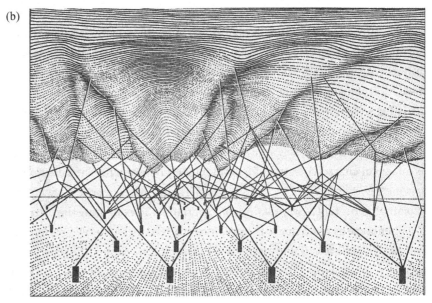

Figure 7.2 Waddington's 'developmental landscape'. Developmental psychobiologists like Gottlieb added non-genetic parameters and feedback loops to this conceptual model, but still treated genetic parameters as the locus of action of abstract developmental genes (e.g. Gottlieb 1992, 186). (Printed with permission from Ruskin House: Conrad H. Waddington, 1957. *The Strategy of the Genes: A Discussion of Some Aspects of Theoretical Biology.* London: Ruskin House/George Allen & Unwin, 29, 36.)

of developmental variables in the population from which it is derived and in which it can be legitimately extrapolated. Thus, the abstract developmental gene explains by reference to the developmental system as a whole, while the Mendelian allele explains by importing statistical information about specific alleles and phenotypes from some reference class (Griffiths and Tabery 2008; Tabery and Griffiths 2010).

From the perspective of the abstract developmental genes it makes no sense to explain the presence of a phenotype (or difference) by alluding to the presence of a particular gene in the absence of any understanding of its role in development. From this perspective the fact that a gene has a specific phenotypic effect raises the question why it has had that effect rather than the other effects it might have had if other variables were different. The claim by developmental scientists that the mere presence of a gene cannot in itself explain the presence of a phenotype reflects this conception of how genes explain phenotypes (Griffiths and Tabery 2008; Tabery and Griffiths 2010).

Conversely, if we conceive of genes as Mendelian alleles, then it will seem unreasonable to demand knowledge about how a gene interacts with other genes and with the environment before accepting an explanation which cites the presence of this allele. If the organism or organisms whose phenotypes are to be explained have been drawn from a suitable reference class, then those other variables will not make a difference (Griffiths and Tabery 2008; Tabery and Griffiths 2010).

As we will see in the next section, an important aspect of the disagreement between behaviour geneticists and developmental psychobiologists has been over the scientific relevance of variables which do not vary in nature. Behaviour geneticists have tended to argue that developmental variables which do not account for any of the variance are irrelevant to the explanation of trait differences. But this overlooks the pattern of explanation associated with the abstract developmental conception of the gene. It is agreed by all sides that the connection between genes and phenotypes proceeds via development. The variables which do not vary are relevant because they are part of the developmental process in virtue of which the variables which do vary exert an influence on the phenotype. Because developmental psychobiology aims to characterise the causal mechanisms of development it has no reason to privilege *actual* difference makers over *potential* difference makers. We will return to this theme at the end of the chapter.

7.3 What is gene–environment interaction?

It is a truism that genes interact with the environment during development. But this truism has been understood in very different ways by developmental scientists and traditional, quantitative behaviour geneticists. In behavioural genetics, interaction is understood as a statistical phenomenon which results in the breakdown in additivity between the genetic and environmental main effects. But in the context of the experimental study of behavioural development interaction is a causal-mechanical phenomenon, not a statistical one. Genetic and environmental factors causally interact in the processes that give rise to phenotypes.

James Tabery has labelled these two concepts of gene–environment interaction the 'biometric' concept ($G \times E_B$) and the 'developmental' concept ($G \times E_D$) (Tabery 2007, 2009). The biometric concept of $G \times E_B$ was introduced by Fisher and $G \times E_B$ became part of the basic conceptual toolkit of quantitative genetics. However, as Tabery has shown, Fisher's formulation of ANOVA was immediately criticised by contemporaries like J. B. S. Haldane and the British biostatistician Lancelot Hogben who thought of gene–environment interaction not merely as a statistical phenomenon produced by the interaction of two sources of variation, G and E, but also as the result of a third causal factor (Tabery 2008). This third factor consisted of the actual, physical combinations of genes and the environment found in the individuals that make up the population. On this view, gene–environment interaction is *manifested* in statistical interaction between measured G and E, but it is *not constituted* by it. Even if no statistical interaction is present in the data, our causal models of development imply that genetic and environmental factors are causally interacting within each organism to produce the phenotype. The failure to observe variance resulting from this causal interaction is something that needs to be explained through an appropriate causal model. Conrad H. Waddington's model of 'developmental canalisation' (Figure 7.2) does just this in the case of the interaction between many genes, by explaining how many different combinations of genes converge on the same developmental outcome (Waddington, 1957).

The longstanding dispute between behaviour geneticists and developmental scientists over gene–environment interaction is to a significant extent the result of their using these two different concepts of interaction. In this dispute, scientists like Gottlieb claimed that gene–environment interaction was ubiquitous despite the failure of behaviour geneticists to detect a large statistical interaction in most studies (Gottlieb 2003, 343). In the same vein, Michael Meaney wrote that '[p]henotype emerges only from the interaction of gene and environment. The search for main effects is a fool's errand. In the context of modern molecular biology, it is a quest that is without credibility' (Meaney 2001a, 51). Behaviour geneticists recognised that two different senses of interaction were in play, but argued that the statistical sense was the only one relevant to population-level studies of individual differences: 'Unfortunately, discussions of genotype–environment interaction have often confused the population concept with that of individual development. It is important at the outset to distinguish genotype–environment interaction from what we shall call *interactionism*, the view that environmental and genetic threads in the fabric of behavior are so tightly interwoven that they are indistinguishable' (Plomin et al. 1977, 309). Tabery has christened this the *defence-by-distinction* (Tabery 2007). It contrasts statistical interaction with 'interactionism', which tries to apply the causal sense of interaction, whose proper domain is the study of individual development, to the study of individual differences in a population.

But Tabery (2009) has argued convincingly that the causal sense of interaction *can* be coherently applied to individual differences in a population. The mechanistic study of behavioural development has traditionally been concerned with the development of species-typical phenotypes, a feature it shares with most traditional developmental biology, but this is not an essential feature of this type of scientific enquiry. Individual differences are as much in need of mechanistic explanation as species-typical phenotypes and in recent years such explanations have started to appear. For example, Michael Meaney and collaborators' work on the molecular basis of individual differences in stress-reactivity (Chapter 5) gives a mechanistic explanation of the distribution of such differences in populations (Meaney 2001b; Weaver et al. 2004). This explanation conceives of gene–environment interaction as $G \times E_D$ and documents how different combinations of gene and environment are distributed in the population. Tabery (2009) has termed the mechanistic explanation of population-level variation the study of 'difference mechanisms'.

The fact that developmental scientists are looking for causal mechanisms explains their interest in variation which does not happen in nature, but which could happen under different conditions. Behaviour geneticists regularly detect large main effects for genes and fail to identify a high level of statistical interaction between genes and environment $(G \times E_B)$. For a behaviour geneticist $G \times E_B$ *is* interaction, and if this element of the variance is low there simply is not much interaction. But for developmental scientists like Gottlieb, interaction is fundamentally a property of causal networks and $G \times E_B$ is only the statistical manifestation of actual causal relationships. Meaney is clearly taking this attitude when he describes his own research and compares it to a statistical approach to the same question:

> The cellular context did not merely determine the magnitude of the glucocorticoid receptor effect on gene transcription, it determined whether that effect was positive or negative – all in relation to a single DNA target. The cellular context, and specifically levels of transcription factors such as cFos and cJun, are heavily influenced by ongoing activity; stress, social encounters – all serve to influence the cellular levels of these factors and can therefore have very potent influences on the nature of gene activity. From such systems will we derive main effects? I think not. (Meaney 2001a, 54)

Meaney suggests that no main effects can be derived from studying the interaction of genes and environment in this system, given that we know that the process is fundamentally interactive. But, of course behaviour geneticists do extract main effects. One response is to suggest that these results are merely a methodological artifact (see Wahlsten and Gottlieb 1997). But another is to recognise that two different conceptions of gene–environment interaction are in play. The absence of $G \times E_B$ is consistent with the presence of $G \times E_D$. The failure to detect statistical interaction tells us either that the relevant variables do not vary in the actual population, or that the developmental system is structured so as to render some developmental outcomes insensitive to variation in those variables (via mechanisms such as redundancy, canalisation, and feedback). Instead of concluding that there is no interaction, we need to find interventions that will reveal it. This explains why developmental scientists like Gottlieb have specialised in using experimental interventions to drive variables to values that would not be encountered in nature. Only by finding such interventions can we decipher the very causal pathways that explain the *lack* of statistical interaction in normal conditions.

These two attitudes to explanation can be related to the discussion of causal explanation in 4.3. Waters (2007) introduced the idea of a specific actual difference maker (SAD) to justify the emphasis on genetic causes seen in certain kinds of scientific work. In Chapter 4 and Chapter 5 we criticised his application of this idea to molecular biology. Waters' other example of the epistemic value of SADs was the way in which classical geneticists like Morgan picked out specific genetic differences between strains as the cause of phenotypic differences. The pattern of reasoning Waters presents with his 'difference principle' (Waters 2007, 558) corresponds to the way in which Mendelian alleles explain phenotypes in behavioural genetics (see 7.2). This is unsurprising considering the close relationship between those two fields. But we also saw in 7.2 that there is another, legitimate kind of explanation which does not make use of the difference principle. Waters says that '[Morgan's] science entailed, as do biological sciences in general, identifying one or a few elements as the "actual cause(s)" in situations that necessarily involve many causes' (Waters 2007, 558–9). But this is an overgeneralisation. There are, indeed, many contexts in which biologists aim to identify salient causes from among the many conditions required for an event to occur. But there are other contexts in which biologists aim to determine how the relationship between variables is mediated by other variables. Developmental science is one of those contexts and, as we have seen, a focus on SADs in this context is precisely the wrong strategy. Instead, the aim is to identify as many of the potential difference makers as possible. Systems biology (Chapter 4) would appear to be another example. This suggests a tentative characterisation of the research contexts that call for this second strategy. They are contexts in which the aim is to discover how to intervene on complex systems (Mitchell 2009).

7.4 The concrete developmental gene

Traditional quantitative genetic methods in behavioural genetics are rapidly giving way to molecular methods, and to the recognition that the understanding of the genetic basis of behaviour means understanding the interaction between multiple genetic and environmental factors (Hamer 2002). The aim of 'gene hunting' is not to identify *the* cause of a behavioural difference, but to provide an entry point to the molecular pathways involved in the development of that difference. Meanwhile, the effects of environmental interventions in developmental science are increasingly being analysed at the level of gene expression (Meaney 2004; Suomi 2004). These developments suggest that we are on the brink of the emergence of a genuinely developmental behaviour genetics that will meet the aspirations of both sides of the earlier debate.

The ground on which the two approaches will meet are concrete sequences of DNA nucleotides. These make concrete the abstract developmental gene, and also constitute the Mendelian alleles whose existence was inferred from quantitative genetic studies. However, the nature of the genetic architecture that emerges from tracking behaviour back to the genome makes this very far from a triumph of reductionism. The science that will be needed to understand how a large number of sequence variations interact with a large number of environmental factors, a relationship undoubtedly mediated by the kinds of epigenetic mechanisms described in Chapter 4 and Chapter 5, will be an example of the 'systems biology' described in 4.7. It will be an example of mechanistic anti-reductionism, using an exhaustive catalogue of parts to understand how an integrated mechanism gives rise to phenomena for which the organisation and the dynamics of the mechanism are key explanatory factors. The explanatory strategy of such a science will involve characterising the whole interactive causal networks with the aim of locating effective points of intervention.

Further reading

A thorough and balanced introduction to behaviour genetics is Kenneth Schaffner's forthcoming book *Behaving: What's Genetic, What's Not, and Why Should We Care?* (forthcoming; and until then Schaffner 2006a, 2006b). Neven Sesardic and Jonathan Kaplan make strong cases on the opposing sides of the debates discussed in this chapter (Kaplan 2000; Sesardic 2005). Two excellent and readable introductions to the developmental science of behaviour are David Moore's *The Dependent Gene: The Fallacy of 'Nature vs Nurture'* (2001) and Patrick Bateson and Paul Martin's *Design for a Life: How Behavior and Personality Develop* (1999).

8 The evolving genome

8.1 Towards an extended synthesis

'Nothing in biology makes sense except in the light of evolution', wrote Theodosius Dobzhansky, but while '[t]here are no alternatives to evolution as history that can withstand critical examination [...] we are constantly learning new and important facts about evolutionary mechanisms' (Dobzhansky 1973, 129). This is as true today as it was forty years ago and a number of biologists now seem to agree with Sahotra Sarkar that 'much of the received framework of evolution makes no sense in light of molecular biology' (Sarkar 2005, 5).

This chapter asks what implications the developments we analysed earlier in the book have for the mechanisms of evolution. These developments include distributed specificity, the idea that a large range of factors share sequence specificity with coding sequences through their role in the regulation of genes expression, and that many of these factors are designed to relay environmental information to the genome. The developments include the revival of notions of epigenesis and plasticity in developmental biology. They include exogenetic heredity, the idea that many non-genetic resources are passed on across the generations and are employed to reconstruct and modify the life cycle through their role in the regulation of gene expression. They include systems biology, which we have suggested has a distinctive style of genetic explanation. Do these developments necessitate an extension of the conventional, neo-Darwinian theory of evolution, the so-called 'Modern Synthesis'?

An answer will in part depend on one's understanding of what constitutes the Modern Synthesis today. Many practitioners have more or less automatically assimilated new conceptual and methodological developments without being aware to what extent some of these violate the underlying assumptions on which the original synthesis was based. Those assumptions are now more than half a century old, and many of the relevant theories and concepts have undergone major revisions. So the point of the call for an 'extended synthesis' may be as much to think through the implications of changes that have occurred or are occurring as to call for more change.

Recent evolutionary thought is clearly marked by increased recognition of the relevance of developmental processes to evolution, mostly but not exclusively under the heading of evolutionary developmental biology ('evo-devo'; Hall 1992, 1998). It is widely agreed that this is an extension of the Modern Synthesis. The mid-twentieth-century synthesis represented the integration of several biological disciplines, notably neo-Darwinian evolutionary theory, classical Mendelian genetics, paleontology, systematics, and the descriptive natural history traditions within zoology and botany. The cornerstone of the synthesis was the mathematical formulation of evolutionary theory as population genetics that had emerged in the 1930s (Mayr and Provine 1980). Evolution was now defined as change in gene frequencies in populations, with genetics alone addressing the traditional Darwinian problems of heredity and variation. The firm rejection of rival late-nineteenth-century accounts of evolution, such as orthogenesis, left natural selection as the sole creative force in evolution. Developmental biology, and to a certain extent ecology, were downplayed in the synthesis. The synthesis also downplayed some traditional questions about biological form. Darwin wrote of 'two great laws' in biology: the conditions of existence, or natural selection, and the unity of type – the distinctive morphological structure of each class of organisms manifest in the homologies between them (Darwin 1964 [1859], 206). While unity of type reflected common ancestry, it was also a factor in its own right in evolutionary explanations of particular characters:

> What can be more curious than that the hand of a man, formed for grasping, that of a mole for digging, the leg of the horse, the paddle of the porpoise, and the wing of the bat, should all be constructed on the same pattern, and should include the same bones, in the same relative positions?
>
> [...]
>
> Nothing can be more hopeless than to attempt to explain this similarity of pattern in members of the same class, by utility or by the doctrine of final causes. (Darwin 1964 [1859], 434–5)

Many nineteenth-century biologists saw morphology – the study of the unity of type – as equally important as, or even more important than, the study of natural selection. But in the Modern Synthesis the traditional idea of 'type' was denounced as pre-Darwinian (Mayr 1976 [1959]) and homologies were understood as the signature of past episodes of selection – 'mere residue of ancestry' (Amundson 2005, 239).

Recent findings in developmental molecular biology have given new life to the unity of type. There is an extraordinary degree of conservation of both regulatory gene networks and protein structure and function. All organisms on earth share the basic processes and components of energy production, protein synthesis, metabolism, and membrane construction. All eukaryotes share the basic compartmentalisation of the cell, gene expression processes of transcription and post-transcription such as splicing, and fundamental cell-cycle and signalling systems (Kirschner and Gerhart 2010). Organisms over a wide range of taxa share their 'genetic toolkit for development' (Carroll et al. 2005, Chapter 2). This has led some evolutionary developmental biologists to argue for a picture of 'regulatory evolution' in which a common set of genetic tools underlie all animal design, as we will discuss below (Carroll et al. 2001; Carroll et al. 2005). Other evolutionary developmental biologists have claimed that homologies are not merely points of resemblance, but natural units of form – the building blocks from which organisms are composed (Wagner 1995, 1996). They have argued that biology needs a systematic theory of phenotypes or 'theory of form' if it is to explain evolutionary novelties – characters which are not modified forms of an earlier phenotypic character, such as the shell of the turtle (Müller 1990, 2003, 2010).

Those who advocate an 'extended synthesis' believe that it will provide more adequate answers to a range of evolutionary questions (Pigliucci and Kaplan 2006; Pigliucci 2007; Pigliucci 2009; Pigliucci and Müller 2010a; Pigliucci and Müller 2010b; Craig 2010). These questions include:

(a) the origins of novel traits (e.g. developmental mechanisms as sources of variation and innovation);
(b) the spread of traits (e.g. co-construction of a selective environment);
(c) the modification of trait (e.g. environmentally induced variability via parental effects);
(d) the pathways that reliably (re)produce traits (e.g. extended inheritance mechanisms). (Stotz 2010; Pigliucci and Kaplan 2006, 128)

In this chapter we do not aim to provide an exhaustive review of these proposals for an extended synthesis, but only to demonstrate the points of contact between them and the ideas in genetics and molecular biology discussed in earlier chapters.

8.2 Developmental plasticity, ecology, and evolution

There has been sporadic interest since the late nineteenth century in the influence of developmental plasticity on evolution (Weber and Depew 2003). There is abundant evidence that organisms can phenotypically accommodate to the environment, and this process could set the stage for further adaptive evolution through *genetic assimilation* (Waddington 1953a, 1953b) or *genetic accommodation* (West-Eberhard 2003). These processes lead to evolutionary change through 'cross-generational changes in frequency distribution of environmentally induced phenotypes' (Badyaev 2009, 1138). Organisms can also use plasticity to buffer themselves against genetic perturbation, a process that may allow for the accumulation of hidden genetic variation that can become visible – and even useful – when environmental conditions change. All of these processes of developmental plasticity could increase a lineage's 'evolvability', its capacity for evolutionary change.

8.2.1 Early evolutionary perspectives on plasticity

Despite the relative neglect of developmental biology by the Modern Synthesis, the mid-twentieth century saw some pioneering work on the integration of development into evolutionary theory. Two prominent examples are Waddington's concepts of 'epigenetic landscape', 'canalisation', and 'genetic assimilation', and Ivan Ivanovich Schmalhausen's concepts of 'stabilising selection' and 'autonomisation' (Schmalhausen 1949; Waddington 1957). Less well known is J. B. S. Haldane and Helen Spurway's (1954) account of transitions between instinct and learning. All three emphasised the reciprocal relationship between ontogeny and phylogeny in adaptive evolution.

From today's perspective Waddington and Schmalhausen are important for their focus on the developmental integration of genetic factors with each other and the interaction of that system with the developmental environment. Both authors envisioned a similar process by which a phenotype that was originally induced by environmental factors could be genetically fixed through selection. Schmalhausen's 'stabilising selection' was selection on novel phenotypes produced by development plasticity operating in environments at the limit of the organism's tolerance. Schmalhausen placed the 'norm of reaction' (Figure 7.1) at the heart of his vision of evolution. A genotype corresponds to a range of phenotypes produced across a range of environments. Under normal conditions development will compensate for environmental fluctuations to produce a relatively constant relationship between genotype and phenotype, but in unfamiliar environments the organism will reveal previously unknown portions of the norm of reaction. The appearance of these novel phenotypes will depend on some specific interaction with the new environment. But if that factor is part of a fluctuating environment, there will be selection for the ability to produce an advantageous phenotype without relying on the environmental trigger ('autonomisation'). Although Schmalhausen's work was endorsed by figures like Dobzhansky, it remained marginal in mainstream evolutionary thought. The term 'stabilising selection' may have been assimilated into the Modern Synthesis, but it became devoid of any developmental connotations.

Waddington introduced similar ideas independently under the name 'genetic assimilation'. He demonstrated experimentally in *Drosophila* that selection for the ability to produce a phenotype in response to an environmental trigger led to the evolution of strains of fly that produced the phenotype without the need for the environmental trigger (Waddington 1953a). In the very same issue of the journal *Evolution* George Gaylord Simpson (1953), the leading paleontologist of the Modern Synthesis, argued that processes such as these add nothing to conventional natural selection. Developmental plasticity simply keeps the organism alive for long enough for conventional natural selection to produce an adaptive phenotype by acting on genetic mutation and recombination. The fact that the plastic response and the subsequent mutations produce the same phenotype is coincidental. Waddington's response was that there is a connection between the developmental response to the environmental trigger and the occurrence of mutations upon which natural selection will act to produce the trait without that trigger:

> I argued that natural selection for the ability to produce an adaptive phenotype would change the genotype in such a way as to encourage the appearance of genetically controlled variance mimicking the adaptive type. The initial non-hereditary response therefore does not merely allow the organism to persist in a new environment and become adapted to it; it enables natural selection to set the stage in such a way that the useful genetic effect is likely to occur. (Waddington 1953b, 386)

This process, while not actually Lamarckian, is still quite different from mere coincidence:

> By speaking of mutations as 'random,' which is true enough at the level of the gene as a protein-DNA complex, we obscure the fact that the effect of a mutation, as far as natural selection is concerned, is conditioned by the way it modifies the reaction with the environment of a genotype which has already been selected on the basis of its response to that environment. This is not neo-Lamarckism, but it is a point which has been unduly neglected by neo-Darwinism. (Waddington 1953b, 386)

The connection that Waddington envisages is forged through the developmental structure of the organism. The developmental mechanisms that make the plastic response possible are the very mechanisms that selection acts upon to produce the genetically assimilated version of the response. The fact that it is relatively easy to select for mutations or gene combinations which produce the phenotype is explained by the fact that the organism has the developmental capacity to produce that phenotype in response to an environmental trigger.

Similar ideas can be found in the less well-known work of Haldane and Spurway on behavioural development (Griffiths 2004). They argue that whether a behaviour is 'innate' or 'acquired' is a superficial feature. What is important is that development is competent to produce the phenotype by some means or other. The genetic basis of an innate and an acquired behaviour, they argue, can be basically the same. The same genetic machinery is activated by different triggers. Consequently they expect to see evolutionary transitions between innate and acquired in both directions, both genetic assimilation and 'genetic accommodation' (see 8.2.2):

> The number of generations during which a learned ethogenesis [*developmental pathway for a behavior*] evolves into an instinctive ethogenesis, if it does so at all, depends on the relative strength of the selection pressures favouring uniformity and variability in development. (Haldane and Spurway 1954, 275)

The key to understanding genetic assimilation is to be aware that becoming autonomous from environmental influences is not synonymous with becoming dependent on genes rather than the environment. The phenotype was always caused by the interaction of a network of genetic and environmental

factors; only now it can be induced by a wider range of values of some environmental factors instead of a very specific value: genetic assimilation is simply
a change in the shape of the norm of reaction (Figure 7.1).

In Waddington's vision of development, the entire collection of genes
makes up a dynamical developmental system which produces a phenotype in
interaction with the external environment (Figure 7.2). The important thing
about a developmentally canalised phenotype is not that it is insensitive to
environmental variation, as is often said, but that there are many exact developmental trajectories that lead to the same place. The primary developmental
processes that underlie a canalised phenotype do not need to change when it
is rendered independent of an environmental stimulus.

The dispute between Waddington and Simpson involves the two different
patterns of genetic explanation described in Chapter 7. Waddington's genes
are what we called there 'abstract developmental genes', the postulated hereditary determinants of parameters in a development model. Abstract developmental genes can only explain a phenotype via the mediation of many other
developmental parameters. In contrast, the Mendelian allele for a phenotypic difference, which was almost certainly what Simpson would have had
in mind, explains that difference without reference to other developmental
parameters. Here lies the reason for the disagreement between Waddington
and Simpson. For Simpson, the genes which do most of the work in explaining
the phenotype are the Mendelian alleles which account for the actual variance
in the trait. These are rare or non-existent when the phenotype is produced
by developmental plasticity. They occur by mutation and are selected because
they are advantageous in the new environment. The two phases of genetic
assimilation thus appear to be unconnected. In Waddington's vision the genes
which do most of the work in explaining the phenotype are the same before
and after genetic assimilation. So the fact that the phenotype can be produced
beforehand by developmental plasticity is explanatorily relevant to the fact
that it can be produced afterwards without any special environmental input.

Waddington and Schmalhausen set out to explain how the plastic response
of an organism could set a direction for adaptive evolution, thus giving the
appearance of the inheritance of an acquired character, and to explain this
in a manner compatible with neo-Darwinism and Mendelism. The conventional Modern Synthesis view was, and still is, that '[d]ifferences due to nature
are likely to be inherited whereas those due to nurture are not; evolutionary changes are changes in nature, not nurture' (Maynard Smith 2000, 189).

The processes suggested by Waddington, Schmalhausen, and Haldane and Spurway were designed to provide a way around this stricture.

8.2.2 Recent evolutionary perspectives on plasticity

In the past twenty years many researchers have followed Waddington and Schmalhausen's lead in investigating the evolutionary impact of organisms' ability to change their phenotype in reaction to changing environmental conditions (Gottlieb 1992; Gottlieb 1997; Schlichting and Pigliucci 1998; Bateson and Martin 1999; Pigliucci 2001; West-Eberhard 2003; Gluckman et al. 2009; Bateson and Gluckman 2011).

One of the best-known evolutionary theorists working on this topic is Mary Jane West-Eberhard. In her view, the various examples of developmental plasticity 'discussed by Kirschner and Gerhart [1998] do not stand as isolated special cases but are part of a larger and more coherent picture of flexible phenotype structure. Their converging views of developmental mechanisms as sources of flexibility that enhance evolvability are likely to have broad application within biology' (West-Eberhard 1998, 8418). West-Eberhard uses a similar argument to Waddington to establish the evolutionary relevance of developmentally produced phenotypic novelties. Novel phenotypic variants, both physiological and behavioural, are produced by developmental plasticity and are then connected with the undirected genetic variation that is almost always present in natural populations to cause evolution. Changes in the frequency of a trait in the population could ultimately be explained by the 'selection on genetic variation in the polygenic regulatory mechanisms influencing its threshold of expression [...] Although mutation is the ultimate source of this variation, mutation need not be associated with the origin of a particular phenotypic novelty' (8418).

New approaches that call for the investigation of organisms embedded in a developmental environment, such as ecological developmental biology ('eco-devo' or 'eco-evo-devo'; Gilbert 2001; Gilbert and Epel 2009; see 5.5.2) or 'developmental ecology' (West et al. 2003), have inspired observations and experiments documenting the impact of the interaction between development and environment on evolution. Gilbert and Epel summarise the plasticity-driven evolutionary mechanisms invoked by Waddington, Schmalhausen, and West-Eberhard as 'change in governance' or 'heterocyberny' (Gilbert and Epel

2009, 372) Control of a developmental process is reassigned from genes to environment or vice versa. In 8.2.1 we described the process of 'genetic assimilation'. 'Genetic accommodation' is the flipside of genetic assimilation, a process by which a phenotype can become *more* responsive to environmental conditions. What is selected for is a plastic rather than a canalised pheno-type, with the ability to react to a range of environmental parameters with different phenotypes, rather than buffer one preferred phenotypic expres-sion against a range of environmental parameters. We saw above that such processes were hypothesised by Haldane and Spurway in the 1950s. Their existence has recently been supported by experimental evidence (Suzuki and Nijhout 2006; Braendle and Flatt 2006).

West-Eberhard (2005b) has also defended the evolutionary significance of a third process, not unlike that which Simpson christened the 'Baldwin effect' and which he regarded as having relatively little theoretical interest (Simp-son 1953). 'Phenotypic accommodation' describes an adaptive developmental response to an environmental or developmental input that is not accompanied by any genetic change. It can be seen as the first step – a developmental plastic response to a perturbation before this developmental response becomes genet-ically stabilised – in the processes of genetic assimilation/accommodation. This 'capacity of organismal homeostasis to accommodate and direct a novel input enabling survival in a novel environment' (Badyaev 2009, 1137) is one of the necessary prerequisites for genetic accommodation to take place and therefore can be regarded as an important process in the production of evolutionary nov-elty. According to West-Eberhard (2005b) phenotypic accommodation should therefore be regarded as an important first step in a process of evolution by natural selection.

Some of these authors have drawn connections between developmental plasticity and 'niche construction'. It was Richard Lewontin who first argued that the selecting environment is not independent of the organism, because organisms choose, combine, and modify the features of the environment with which they interact, and thereby construct their own niche: 'Organisms fit the world so well because they have constructed it' (Lewontin 1996, 10; see also Lewontin 1983a, 1983b). This idea has been developed into the theory of 'niche construction' (Odling-Smee 1988; Odling-Smee et al. 2003). Through niche construction populations influence their own selection pressures and the feedback between natural selection and niche construction can change the dynamics of the evolutionary process. Niche construction theory stays close to the received view according to which environments are the agents that select (rather than induce) variations, but it still gives the developmental system a more active role in evolution.

The impact of these processes on evolution is not yet clear, but some of their advocates think it very substantial. West-Eberhard suggests that 'genes are probably more often followers than leaders in evolutionary change' (West-Eberhard 2005a, 6543).

8.3 Exogenetic inheritance

In 5.5.3 we encountered the idea of exogenetic inheritance, causal pathways by which parents can influence offspring phenotypes other than via nuclear DNA. This is a very different conception of heredity from that associated with the Modern Synthesis. It has been argued that it is a return to a way of thinking about heredity that existed before Mendelian genetics.

According to Ronald Amundson the early twentieth century saw a reconfiguration of the idea of heredity which radically separated it from development. The inheritance of a Mendelian allele explains the inheritance of a phenotypic trait, but it does so without explaining how that trait develops. The allele causes a certain character, say the colour of flowers:

> [We] may say that a particular factor (p) is the cause of pink, for we use cause here in the sense in which science always uses this expression, namely to mean that a particular system differs from another system only in one special factor. (Morgan et al. 1915, 209; cited in Amundson 2005, 149)

Morgan's idea of 'cause' did not require knowledge of the underlying mechanism by which the inheritance of the allele produces the phenotype. It is the idea of causes as actual difference makers described by Waters (see Chapter 4), and which we also encountered in our discussion of traditional behaviour genetics (Chapter 7).

However, to earlier biologists the reliable reappearance of a trait in the next generation called for an explanation in terms of its ontogenetic history. Accordingly, it seemed only natural to understand heredity in terms of development:

Indeed, heredity is not a peculiar or unique principle for it is only similarity of growth and differentiation in successive generations [...] The causes of heredity are thus reduced to the causes of successive differentiation of development, and the mechanism of heredity is merely the mechanism of differentiation. (Conklin 1908, 90; cited in Amundson 2005, 148)

Understanding heredity as the inheritance of actual difference makers separates questions about heredity from questions about development – about how those difference makers produce their effects. However, to complete the Modern Synthesis picture of evolution as change in gene frequencies, it was necessary to exclude other difference-making factors from the theory of heredity. The relevant criterion was, and continues to be, the instability of what was termed 'soft inheritance'. Phenotypic changes due to the environment are either not inherited at all, or are too unstable and fluctuating to be the basis of cumulative change by natural selection. Since it is cumulative selection, rather than one-step selection, that produces complex adaptation this would seem to relegate exogenetic inheritance to a minor role at best. The key implication of this is that the influence of the environment on phenotypes can only affect the course of evolution if it can be somehow written into the genes. This explains the recurrent interest in genetic assimilation as well as the short-lived excitement about the discovery of reverse transcription from RNA to DNA (see Chapter 3).

Advocates of extended heredity have two possible replies. First, they can argue that other forms of heredity are not as unstable as normally supposed. Second, they can question whether the evolutionary significance of an inheritance system really turns on its stability across the generations. Eva Jablonka and her collaborators have made significant efforts to mount the first kind of defence. They have documented the surprising extent of behavioural and cultural inheritance across a wide range of animals (Jablonka and Avital 2001), and produced a review which reveals that epigenetic inheritance in the narrow sense – the transmission of gene expression patterns through meiosis – can last up to three generations in humans and up to eight in other taxa (Jablonka and Raz 2009).

But even if exogenetic inheritance is more stable than is normally supposed, this may not mean that it can play a similar evolutionary role to genetic inheritance, generating the kind of variation on which natural selection can feed to produce complex adaptations. In an important challenge to the idea of extended heredity Maynard Smith and Eörs Szathmáry (1995) argued that the genetic inheritance system and cultural transmission in humans are the only two systems that display what they call 'unlimited heredity', the form of heredity that makes possible the evolution of complex adaptation. Most inheritance systems can only mutate between a limited number of heritable states which can be specified in advance. DNA methylation, for example, can only choose whether existing genes will be switched on or off. The genome and language, however, both have a recursive syntactic structure. Their basic constituents can be put together in many different combinations and these combinations can be of any length. Hence these inheritance systems have an unlimited number of possible heritable states. We are sympathetic to this idea, which can be used to help us to understand why the informational specificity of the genetic heredity system constitutes a key innovation for the transmission of biological specificity between cells. But we do not think the distinction should be used to assess the evolutionary significance of heredity systems more generally.

The limited/unlimited heredity distinction overlooks the fact that all inheritance systems operate at the level of the whole developmental systems (Jablonka 2001, 100; Griffiths 2003). Maynard Smith and Szathmáry's argument assumes that the number of permutations of RNA codons is the relevant measure of 'limitedness' for the genetic inheritance system and that some corresponding measure of the number of permutations of physical parts is the appropriate measure for other inheritance systems. But for any one inheritance system, the physical changes that count as changes from an evolutionary point of view are only those which appear as changes to the developmental system as a whole. This point can be seen by looking at human languages. Not all physical differences between syntactic objects are *syntactic* differences. Differences in handwriting, for example, are not syntactic differences. Syntactic differences are only those physical differences that count as differences to the broader system that uses the physical inscriptions. The same phenomenon occurs in the genetic heredity system. The existing genetic code is substantially redundant, with more than one codon corresponding to the same amino acid (Table 3.1). That it is not more redundant is explained by the evolved coding scheme – the genetic code. What that actually means is that the cellular machinery treats most physically distinct codons as different signals. The genetic code is not a fact about DNA, but a fact about the machinery of translation, especially the population of tRNAs in the cell. Generalising this point, it becomes clear that changes in the developmental systems can expand the possible interpretations of existing developmental resources, including genes. The evolutionary possibilities that can be caused by a nucleotide substitution in the DNA of a eukaryote cell could not be caused by those same substitutions in a prokaryote cell. Hence, the unlimited nature of the genetic inheritance system is more accurately seen as a property of the developmental system as a whole and not of the genome in isolation.

In effect, the measure of 'limitedness' used by Maynard Smith and Szathmáry is biased. It allots to the genetic inheritance system all the outcomes that can be generated by making changes to that system across the full range of possibilities for exogenetic inheritance, while allocating to exogenetic only the number of outcomes it could produce given one possible genome. An advocate of extended heredity could reply in kind by allocating to the genetic inheritance system only the range of outcomes it could generate given one state of exogenetic inheritance systems and allocating to an exogenetic inheritance system all the outcomes it could generate given the whole range of possible genomes. Both of these moves ignore the fact that heredity is produced by the interaction of these systems. The limited/unlimited heredity distinction may show that epigenetic inheritance systems could not have evolutionary significance in the absence of the genetic inheritance system, a plausible thesis since nucleic acid-based heredity is clearly a key innovation in the history of life. But it does not show that epigenetic heredity systems are of little evolutionary significance.

Another reason not to accept that epigenetic inheritance systems must be stable across many generations to be of evolutionary significance is that their evolutionary significance may lie precisely in their relative instability. Because organisms have to cope with fluctuating environments as well as stable ones, inheritance systems may serve their collective purpose best if they are situated on a continuum between stability at one end and inducibility at the other (Lamm and Jablonka 2008; Badyaev 2009).

Finally, we would reiterate the point we made in our discussion of extragenetic inheritance in Chapter 5. The role of a theory of heredity in evolutionary theory is to specify how the phenotypes of parents are related to the phenotypes of offspring. This is the role played by a Mendelian account of heredity in conventional population genetics and quantitative genetics. So in one very important sense of 'evolutionary significance' we already know that extragenetic inheritance is significant: if it is left out of an evolutionary model then the model will give inaccurate predictions about the evolutionary trajectory of the population. This point is already well appreciated in the population genetics literature on parental effects (Wade 1998).

8.4 Variation, regulatory evolution, and the origin of novelty

The architects of the synthesis thought that variation is independent of the environment (and therefore not adaptively directed), unconstrained and uniform in all directions, abundant, and has small effects on the phenotype. These assumptions about the nature of variation were not based on strong empirical evidence at the time, and were not confirmed with the development of molecular biology and developmental genetics. The assumptions flowed over into another basic assumption that the Modern Synthesis inherited directly from Darwin: evolutionary change happens gradually, an assumption that has been regularly challenged (Pigliucci and Müller 2010b).

The most radical challenge to these assumptions concerning variation comes from mechanisms whose effect is equivalent to the inheritance of acquired characters, some of which we discussed in 8.3. But challenges have also come from evolutionary developmental biology. Despite its many departures from the Modern Synthesis view of evolution, evo-devo has stayed a gene-centred field, with most practitioners maintaining that only genetic resources are inherited (Robert et al. 2001). Because evolutionary developmental biologists are interested in phenotypic evolution this made it important for them to understand the so-called 'genotype–phenotype map'. For the architects of the synthesis the concept of the 'genetic programme' had disposed of this problem (Gehring 1985). Evo-devo has revived the study of the actual mechanisms that connect genetic change to phenotypic change, and with this has come the possibility that phenotypic variation may be highly constrained, far from uniform in direction, and large as well as small.

8.4.1 Regulatory evolution

In the 1970s molecular biology discovered so-called 'regulatory genes', genes coding for products which regulate other genes, as distinct from structural genes coding for the building blocks of the cell, and also the existence of regulatory information outside coding sequences (e.g. *cis*-regulatory modules and enhancers; see Chapter 4). This led to the development of the first models of gene regulatory networks (GRNs) (Davidson 2001; Davidson and Levine 2005; Davidson and Erwin 2006). More recently it has been the basis of a new view about evolution, the 'cis-regulatory hypothesis'. This claims that evolutionary change depends less on small changes in the coding region of genes than on the way structural genes are regulated (Carroll 2005; Stern and Orgogozo 2008; Pennisi 2008).

The scientists behind the cis-regulatory hypothesis maintain that to understand the evolution of complex body plans requires understanding the 'logic processing system' of the regulatory genome. Classic evolutionary theory has been unable to provide 'an explanation of evolution in terms of mechanistic changes in the genetic regulatory program for development of the body plan, where it must lie' (Davidson and Erwin 2006, 796). Sean Carroll has described the 'primacy of regulatory evolution' based on his own account of the developmental toolkit that lies behind the ontogeny of the body plan of animals, and the opportunities it offers for changes in the regulatory architecture as the evolutionary material for variation (Carroll et al. 2001; Carroll 2005; Carroll et al. 2005).

In 4.4.1 we described some of the processes involved in the regulation of genome transcription, and mentioned that 50–100 proteins, including chromatin-modifying proteins, transcription factors, and their co-factors, are typically involved in the transcription of a gene in a multi-cellular eukaryote. So it should come as no surprise that some of the best-known and most ancient gene families code for these regulatory proteins. Gene families are classified according to some functional protein domains common to all family members; in transcription factors that is typically their DNA binding domain. The best-known family among the transcription factors is the Hox genes. These transcription factors belong to a developmental regulatory system that orchestrates the positional identities of cells along the anterior–posterior axis. The 'homeobox' domain unites all Hox proteins. They were first detected in *Drosophila* when mutations in these genes created what William Bateson had earlier named homeotic transformations (Bateson 1894). These are mutations which replace one body part – say, a wing – with another – say, a forelimb – or which cause the duplication of a body part. *Drosophila*, like other insects, has two Hox gene clusters, bithorax and antennapedia, comprising eight different Hox genes which are believed to be derived from single gene duplications. Later studies revealed genes with a very similar homeodomain in a wide range of animals. Hox genes in all taxa are structurally, biochemically, and functionally conserved (Ruddle et al. 1999; Carroll et al. 2001; Carroll et al. 2005). For proponents of evolutionary developmental biology, Hox genes and the so-called 'genetic toolkit' for development which they embody are of fundamental importance in understanding evolution. In their role of 'master control genes' they are thought to give insight into how changes at the genetic level could lead to major changes at the phenotypic levels. They could also explain the origin of morphological novelties. Carroll and his colleagues have argued that these genes are the creative force that underlies the morphological diversity of animals (Carroll et al. 2001, 213f.).

Philosopher Jason Scott Robert (2001) has criticised the idea of 'master control genes'. It would be a mistake to think that the dramatic effects of Hox genes are to be explained by those DNA sequences in themselves. Their ability to do what they do is conferred on them by their role in the developmental system. Waddington's ideas outlined in Chapter 7 and 8.2 are helpful here. The effects of a change in a Hox gene is an example of the theoretical phenomena Waddington described as switching development from one canalised pathway to another. In fact, some of the phenotypes he studied were produced by mutations in Hox genes. So these genes are an example of how abstract developmental genes can be made concrete by molecular biology.

8.4.2 The origin of form

Developmental accounts of evolution have criticised the answers given by the Modern Synthesis to evolutionary questions, but they have also identified questions that the architects of the Modern Synthesis marginalised or ignored. Philosophers Alan Love and Ingo Brigandt have argued that two such questions, the origin of morphological novelty and the origin of functional innovation, are at the centre of the explanatory agenda of evolutionary developmental biology (Love 2001; Love 2008; Brigandt and Love 2010).

Darwin addressed the 'survival of the fittest', but as some of his contemporaries complained, he did not address the 'origin of the fittest' (Cope 1887). What are the developmental origins of the variation on which natural selection acts? The Modern Synthesis did not regard this as a substantive issue – small variation is ubiquitous in all directions and selection moves organisms gradually from one form to another. But evolutionary developmental biology has revived this question of the 'arrival of the fittest' (Fontana and Buss 1994), and the origin of form is at the forefront of questions asked by critics of the synthesis. These critics allege that the Modern Synthesis lacks an account of where phenotypes come from – an 'evolutionary biology of organismic design' (Wagner 1994, 276). Evolutionary developmental biologists Gerd Müller and Stuart Newman titled one article 'Origination of Organismal Form: The Forgotten Cause in Evolutionary Theory' (Müller and Newman 2003; see also Müller and Wagner 1991; Newman and Müller 2000). Müller and Newman think that developmental plasticity is an important part of the solution. In the same vein, parental effects researcher Alexander Badyaev (2011) believes that developmental plasticity, or what he terms 'emergent variation', will contribute to solving the problem of the 'origin of the fittest'. Marc Kirschner and John Gerhart's theory of 'facilitated variation' is also an attempt to resolve what they call 'Darwin's dilemma', his lack of a theory of variation, which 'left a glaring gap in understanding how animals develop their astounding variety and complexity'. More importantly, 'ignorance about novelty is at the heart of skepticism about evolution, and resolving its origin is necessary to complete our understanding of Darwin's theory' (Kirschner and Gerhart 2005, x). For these authors it is the organisation of the development system that facilitates the creation of new variation (Kirschner et al. 2000; Kirschner and Gerhart 2005; Kirschner and Gerhart 2010).

The origin of form has attracted a lot of recent scientific attention, but some advocates of the extended synthesis still doubt that much has been achieved in the way of 'new conceptual advances toward a convincing theory of form' (Pigliucci 2007, 2745).

8.5 Selection and constraint

> The only constraints on evolution are the very systems that make natural selection (via development and the reliable inheritance of developmental systems) possible [. . .] We prefer to think of developmental processes and systems of inheritance as explaining the possibility and reliable replication of adaptations. (Pigliucci and Kaplan 2006, 127)

The evolutionary mechanisms recognised by the Modern Synthesis were the stochastic process of drift, mutation pressure – the fact that some mutations happen more readily in one direction than in another – and natural selection. Natural selection was seen as the only creative force in evolution. Because the organism and its developmental processes were the expression of instructions in the genetic programme, and variations were understood as undirected, any creativity must lie in the sorting process of natural selection.

Since the 1980s another factor, developmental constraint, has moved into the orbit of the Modern Synthesis. Developmental constraints were defined as 'a bias on the production of variant phenotypes or a limitation on phenotypic variability caused by the structure, character, composition, or dynamics of the developmental system' (Maynard Smith et al. 1985, 266). Developmental constraints are commonly presented as constraining natural selection by limiting what variation is available for natural selection to work with. But many theorists have rejected this view and have emphasised the constructive and enabling aspect of constraints – these limit variation in certain directions, but open up the options in a range of others (Alberch 1982; Maynard Smith et al. 1985; Wimsatt 1986b; Juarrero 1998; Amundson 2005). Others have rejected the whole idea that natural selection is constrained by the constitution of developmental systems.

> The fact that organisms are physical entities with complex developmental histories and particular systems of inheritance is not a constraint on the power of natural selection; rather [. . .] it is what makes natural selection possible. Of course, these physical features do influence the kind of adaptations that are likely, as well as how quickly those adaptations will spread, and what their chances are of being maintained. (Pigliucci and Kaplan 2006, 126; see also Kauffman 1993; Weber and Depew 1996)

This disagreement about whether constraints are opposed to natural selection reflects the different roles of the concept of constraint in different areas of research. For an evolutionary ecologist, a developmental constraint is a postulated explanation of why a phenotype that would seem an obvious solution to an adaptive problem does not actually occur. So constraints appear to work against natural selection. But biologists who emphasise the positive role of constraints do so as part of work on the origin of form or the origins of variation. Looked at from this perspective, constraints are not constraints on selection, but constraints on chaos. In studies of self-organising complex, adaptive systems it is the constraints on these systems which are the source of order (Kauffman 1993; Goodwin 1994; Weber and Depew 1996; Morowitz 2003). Constraints can be thought of as boundary conditions. Living systems build richly interwoven webs of boundary conditions that constrain the release of energy to allow them to maintain themselves in highly ordered states far from thermodynamic equilibrium.

The ideas, concepts, and theories derived from complexity theory and theories of self-organisation have not yet had a great deal of uptake in conventional evolutionary theory. But some of those calling for an extended synthesis, such as Massimo Pigliucci (2007), see potential value in this approach. Some forms of self-organisation could also explain the very earliest origins of form, before natural selection appeared on the stage (Weber and Depew 1996; Newman 2003; Newman 2010).

8.6 An extended synthesis?

Evolutionary developmental biology, ecological developmental biology, and some of the other research agendas discussed here represent a shift in focus from population-dynamic models to causal mechanistic accounts of the processes by which evolutionary change occurs. They suggest that, *contra* Mayr, the answers to proximate questions are of relevance to evolutionary questions (Laland et al. 2011). Answering those proximate questions, mostly about development, would allow an extended synthesis to address the full range of *how* questions concerning the origination, fixation, modification, and reconstruction of characters.

We have described how Waddington's idea of genetic assimilation depended on integrating an account of how the phenotype develops with an account of its evolution. With the revival of interest in integrating development and evolution, genetic assimilation and genetic accommodation are now firmly back on the evolutionary agenda. But they represent only two of many ways in which the interaction between development and the environment may impact on evolution. We noted that interest in these two processes is partly driven by the idea that if the interaction between development and the environment is to be of any evolutionary significance, its results must somehow be written into the genes. In 8.3 we argued against this idea and for the evolutionary importance of extragenetic inheritance in its own right. We described the way in which the Modern Synthesis approach to heredity tried, and still tries, to exclude extragenetic inheritance from evolutionary significance, and argued against this. Even if extragenetic inheritance were as unstable, fluctuating, and 'limited' as its critics allege, it would still have a significant evolutionary impact.

In 8.4 and 8.5 we looked at some of the more radical research agendas that have grown out of and around evolutionary developmental biology. These involve changing the questions asked by evolutionary biologists as well as improving the answers to existing questions. The unity of type and the origins of form are back on the explanatory agenda of evolutionary theory. While some research on these questions remains focused on the genome, other proposals draw on the ideas about plasticity and the role of the environment in development which we addressed above. Finally, we briefly mentioned work that looks to the dynamics of development and self-organisation as an additional source of order in living systems that can complement natural selection.

A synthesis between all the new and old strands of evolutionary thought will not be worked out in a few papers, book chapters, or conferences. There is no consensus yet about what the final form of any extended synthesis will look like: what the explanatory foci will be, what contributing disciplines and areas, what assumptions need to go for good and what new ones will replace them. But whatever the final form of the new synthesis, it will represent progress for evolutionary theory as it enlarges its explanatory scope.

Further reading

The obvious place to find out more about these ideas is the volume of essays *Evolution – The Extended Synthesis* (Pigliucci and Müller 2010). Massimo Pigliucci and Jonathan Kaplan criticise many of the assumptions of conventional, Modern Synthesis Darwinism in *Making Sense of Evolution* (2006). *The Transformations of Lamarckism* is a good collection of essays on efforts to link phenotypic plasticity and evolution (Gissis and Jablonka 2011). *From DNA to Diversity* is a very accessible introduction to the 'genetic toolkit' approach to evolution (Carroll et al. 2005). *The Plausibility of Life* is a radical challenge to the assumptions about variation with which we opened this chapter (Kirschner and Gerhart 2005).

9   Four conclusions

9.1 The identities of the gene

The gene today has several identities which have accumulated as the molecular biosciences have developed and diversified. The gene is still an instrumental unit for genetic analysis – the Mendelian allele. The gene is also a material unit of heredity, a reasonably clearly defined structural unit used in annotating genomes – the nominal gene. The gene is also a unit of Crick information, a volume in the 'library of specificities' (Nanney 1958). But the relationship between this identity of the gene and its identity as a structural unit has become increasingly vexed in recent years. So the gene as a unit of Crick information has taken on a separate identity – the postgenomic gene. The gene also has less prominent identities: we saw that some 'genes' are no more than hypothesised anchors for the parameters of developmental models – 'abstract developmental genes'. Each of these identities plays a productive role in some forms of biological research. Scientists are adept at thinking about genes in whichever way best suits their work, and at switching between these different representations of the gene as the nature of their work changes. The concept of the gene is therefore best thought of as a set of contextually activated representations.

In Chapter 2 we described how the gene of classical, Mendelian genetics had two identities, as an instrumental unit for genetic analysis and as a hypothetical material unit of heredity. In Chapter 3 we described the elucidation of the basic structure and function of DNA. This represented the successful conclusion of the search for the gene as a material unit of heredity. However, the causal role of the gene as it had been envisaged in classical genetics was very substantially revised in order to fit what had been discovered about the material basis of heredity. Moreover, the original role of the Mendelian gene continues to define the gene whenever biological research uses genetic analysis. As we showed in Chapter 3, it can be necessary to think of genes as both Mendelian alleles and molecular genes, even when those two identities do not converge on the same pieces of DNA. The Mendelian gene turned out to be grounded in DNA, but the development of genetics has left us with more than one scientifically productive way of thinking about DNA and the genes it contains.

The development of the gene concept does not fit conventional philosophical models of the evolution of theoretical terms, particularly models which presume that conceptual evolution takes the form of growing knowledge about a single entity – 'the gene'. When Mendel talked of 'factors' and Johannsen introduced the term 'gene' they were not simply referring without knowing it to the molecular gene.[1] Nor was the result of the molecular revolution in genetics that a previously defined causal role, the Mendelian gene, was filled by a newly discovered material occupant, the molecular gene. Instead, various uses of the term 'gene' relate to different kinds of research, and different representations of the gene feature in those forms of research. The molecular gene could only replace the Mendelian gene if it could play the same role in genetic analysis. However, there are many pieces of DNA which are not molecular genes but which still yield to genetic analysis: they function as alleles that occupy loci. The molecular gene represents only one way to fill the role of the Mendelian gene. Conversely, the molecular gene is anchored in its own experimental practice and cannot be redefined to apply to any piece of DNA which can act as a Mendelian gene. The birth of molecular biology introduced a new research focus on the structural analysis of genes and the elucidation of their relationship to their products. The gene in its molecular guise acquired a new function – the synthesis of a gene product – as well as a definite structure. Research into this new function was based on the linear correspondence between genes and their products. To play its role the molecular gene must be a sequence which can be seen as the linear image of a gene product.

So rather than seeing the Mendelian gene as a primitive precursor to the molecular gene, we think a better account is one which recognises that the gene always had two identities, and that it retains those identities even today. There is no ontological mystery here – both identities are anchored in facts

---

[1] For an attempt to develop a more sophisticated version of the 'causal theory' of reference for theoretical terms to handle complexities like that of the gene concept, see Stanford and Kitcher (2000).

about DNA. Each of these ways of thinking is grounded in real facts about what the underlying molecules are doing, but it is not possible to reduce them all to one uniform way of dividing the genome into genes.

In Chapter 6 we examined another identity of the gene, as a unit of information. The idea that genes are instructions for phenotypes, imposing form on matter in a semantic version of the preformation theory, results from confusing evolutionary explanations of why development occurs with mechanistic explanations of how it occurs. The search for the semantic gene has led philosophers to misinterpret what models derived from computing and information theory are doing in the molecular biosciences. However, we concluded that in reacting against this mistake, critics such as ourselves have paid too little attention to less overblown but very important informational ideas in biology. The genetic code is a means to transfer information in the sense defined by Francis Crick (1958, 1970), the causal specification of the sequence of a gene product, or Crick information. Crick information is not contained solely in nucleic acid sequence, as we showed in Chapter 4 and Chapter 5 and reiterate below, but the ability of organisms to transfer specificity between generations is crucially dependent on the invention of nucleic acid-based heredity.

The conception of the gene as a unit of Crick information corresponds to what we called in Chapter 4 the postgenomic gene. The processes of genome regulation which we discussed in Chapter 4 and Chapter 5 mean that, rather than containing a determinate number of genes with a determinate function, the genome contains a range of coding sequences, not all of which lie within 'nominal genes', which are used in a flexible, context-dependent way to make gene products. The continuing focus of molecular biology on determining how the genome makes its products requires a flexible, top-down conception of the gene as any set of coding sequences that can be used as the templates for a gene product. It is in this context that genes take on their postgenomic identity as 'things you can do with your genome'.

In Chapter 7 we introduced one more identity for the gene, the 'abstract developmental gene'. Genes take on this identity in contexts where scientists conceive of genes as mechanistic causes of development, but are unable to characterise them at the molecular level. This causes genes to appear in an extremely abstract guise as the determinants of the value of parameters of a developmental model. In this chapter we saw the importance of the fact that the identities of the gene are all ultimately grounded in facts about DNA. We described some longstanding disputes between developmental psychobiologists, who until recently have used abstract developmental genes in their models, and behaviour geneticists. The latter think of genes as Mendelian alleles, but they do so in the context of quantitative genetic models and until recently have rarely identified the alleles in these models with specific loci or DNA sequences. We showed how intractable disputes between these groups are now starting to evaporate because both fields have 'gone molecular' and can relate their claims to specific DNA sequences. We suggested that this was an example of the value of 'boundary objects' (Star and Griesemer 1998) which can move between different intellectual contexts. Generalising this example, we tentatively suggest that the availability of well-characterised lines of organisms, of chromosomes, and later of DNA sequences, all of which can act as 'boundary objects' between different research contexts, may explain how genetics has been so successful with such a multi-faceted and contextual object of study as the gene.

9.2 Molecular epigenesis

Another theme that runs through the previous chapters is that recent developments in the molecular biosciences have undermined the idea that genes, however understood, are the prime movers in all biological processes. Despite the key role of nucleic acid inheritance in making it possible to move biological specificity between generations, there is more to heredity than the inheritance of nuclear DNA. Although all biomolecules are ultimately synthesised from nucleic acid templates, those templates are not the only source of the specificity of biomolecules. Finally, despite the importance of gene control networks in the regulatory architecture of the cell, the complete regulatory apparatus includes a much wider 'developmental niche'. It is claims like these that make up the 'parity thesis' (Oyama 1985; Griffiths and Knight 1998; Oyama 2000a; Griffiths and Gray 2005). The specific roles played by the gene in its many guises are more than enough to explain its central place in biology research. There is no need to look for some fundamental, metaphysical difference between genes and the rest of biological reality or some fundamental epistemological difference between genetic explanations and other biological explanations.

A major theme of Chapter 3 was the emergence of 'informational specificity' as the defining property of the molecular gene. The search for the molecular gene was the search for the source of biological specificity – the ability of biomolecules to catalyse very specific chemical reactions. We described how specificity was transformed from a physical concept based on stereochemistry – the three-dimensional shape of molecules – to an informational concept based on the linear correspondence between molecules, most famously in the case of the genetic code. The term information in 'informational specificity' refers to the causal role played by the order of DNA nucleotides in determining the sequence of elements in the molecules derived from it. The order of nucleotides in DNA is a *causally specific difference maker* with respect to the sequence of elements in the product, or the *instructive cause* of that sequence, two ideas discussed in Chapter 4. We introduced the term 'Crick information' to refer to information in this sense.

Chapter 4 explored the dramatic changes that have occurred in the understanding of genomes and their components in the 'postgenomic era'. These developments have challenged the idea that DNA sequences are the only source of Crick information. We described cellular processes that allow cells to make many times more products than they have nominal genes through the activation, selection, and creation of Crick information. We introduced the concepts of 'genetic underdetermination and amplification' and 'distributed specificity' to describe the relationships between the DNA coding sequences, the regulatory machinery that makes use of them, and gene products. We also introduced 'molecular epigenesis', the idea that even the immediate products of genome expression are products of a wider developmental system.

Many biologists familiar with the mechanisms described in Chapter 4 think that their action can be traced back to other regions in the genome, so that the genome as a whole remains the sole source of specificity. In Chapter 5 we argued that this is not the case. Regulatory mechanisms contribute Crick information by acting as causally specific difference makers with respect to the sequence of the product. The information they contribute is not, at least in most cases, the Crick information from the coding sequence from which regulatory molecules are transcribed – the sequence of the final product does not have the right kind of relationship to the sequence of the regulatory molecule. Moreover, regulatory mechanisms are influenced by factors outside the genome which act as causally specific difference makers. Some of these difference-making factors are located in parental genomes, but others are not. Developmental plasticity affords some important examples. Organisms adjust their phenotypes in the light of information from the environment which they themselves or their parents transmit to the genome through its regulatory machinery. Other examples come from the role of exogenetic inheritance in normal development. We introduced the idea of the 'developmental niche' as a good integrative framework for the various ideas about heredity 'outside the genome' surveyed in Chapter 5. The developmental niche contains all the legacies that are required for properly regulated expression of the genome throughout the life cycle of the organism.

9.3 Genetics and reductionism

The story of genetics can be told as one in which the gene was glimpsed only dimly through the dark glass of genetic analysis, and progressively better understood until its true molecular nature was revealed. In the 1960s philosophers suggested that Mendelian genetics was being reduced to a newer, more adequate theory called molecular genetics, just as Newton's theory had been reduced to Einstein's. However, as we explored in Chapter 3, genetics proved intractable for traditional models of theory reduction. These difficulties reflected the multiple identities of the gene. The Mendelian gene turned out to be grounded in DNA, but the development of genetics has left us with more than one scientifically productive way of thinking about DNA and the genes it contains.

Recent philosophical discussion of reductionism has focused not on the relationship between different theories in genetics, but on whether the explanations which molecular biology provides are reductionist explanations. We argued that the way in which molecular biology explains biological phenomena is best captured by the neo-mechanist account of reduction as the elucidation of underlying mechanisms associated with authors such as William Bechtel (2006, 2008). Mechanistic explanations include both a reductionist phase which identifies the constituent parts, and an integrative phase that explains the specific ways in which those parts are organised so as to produce the phenomenon. Molecular biology, we argued, is both reductionist and integrative in this way. The rise of systems biology strongly supports this picture, as we argued in Chapter 4. It retains a commitment to the reductive strategy of decomposing systems into their components and characterising those components, but its explanatory strategies rely on the functional organisation of systems and the top-down effects of that organisation on the interaction between the parts.

Genetic explanations of behaviour, especially human behaviour, are often regarded as reductionistic. In Chapter 7 we saw that this is increasingly inaccurate as behaviour geneticists move from studying statistical relationships between genes and behaviour to uncovering causal mechanisms. Meanwhile, the effects of environmental factors on behaviour are increasingly analysed at the level of gene expression. The science that will be needed to understand how a large number of sequence variations interact with a large number of environmental factors via epigenetic mechanisms of genome regulation to produce behaviour is almost certain to be another example of mechanistic anti-reductionism. It will use an exhaustive catalogue of parts to understand how an integrated mechanism gives rise to phenomena for which the organisation and the dynamics of that mechanism are key explanatory factors. The explanatory strategy of such a science will not focus on a search for single causes, but on characterising whole causal networks.

9.4 Nature and nurture

Much of the material surveyed in this book contributes to the long-drawn-out demise of the nature/nurture dichotomy. Biological thought is haunted by the many incarnations of this dichotomy and the assumptions it creates about the structure of biological systems and the nature of explanation in the biological sciences.

The successes of genetics and molecular biology are often seen as a triumph of nature over nurture. But the developments surveyed here reveal that this is a mistake. In an attempt to uncover the underlying 'nature' of organisms, molecular biology has revealed the interdependence of organism and environment. The regulatory architecture of the genome reaches outside the genome itself, outside the cell, and outside the organism. Factors outside the gene not only activate, they differentially select and they create biological information. The basis of biological specificity is distributed between coding sequences, regulatory machinery, and the broader developmental niche. Many of the factors involved in genome regulation are highly context-sensitive, which allows them to relay environmental information to a reactive genome which has evolved to let environmental inputs play an instructive role in the determination of phenotypes. The overall picture is one of molecular epigenesis.

Meanwhile, new perspectives have developed on nurture. Some of these focus on developmental plasticity – the ability of organisms to modify physiological, morphological or behavioural phenotypes in response to their environments. Others focus on the role of exogenetic processes in biological heredity. Much of this effort has been directed at cellular epigenetic heredity systems, but others have started to investigate inheritance at the behavioural/psychological, ecological, linguistic/symbolic, socio-cultural and cognitive-epistemic levels. Much of the new science of nurture adopts a reductionistic research strategy, tracking both the process of nurture and its effects down to the molecular level.

A more epigenetic understanding of nature together with a more mechanistic understanding of nurture renders many of the old dichotomies blurred or entirely incoherent. Genes, in some of their guises, are defined by their broader context. The environmental is an essential component of the evolved developmental system. Heredity is a mechanism for plasticity as well as fixity. Perhaps most importantly, at least for a book in the philosophy of science, research in the molecular biosciences is both strongly reductionist and strongly integrative.

Bibliography

Alberch, Pere. 1982. Developmental constraints in evolutionary processes. In *Evolution and Development*, ed. J. T. Bonner. New York: Springer Verlag, 313–32.

Alberts, Jeffrey R. 1994. Learning as adaptation of the infant. *Acta Paediatrica* 83 (s397): 77–85.

2008. The nature of nurturant niches in ontogeny. *Philosophical Psychology* 21 (3), special issue, *Reconciling Nature and Nurture in the Study of Cognition and Behavior*: 295–303.

Amundson, Ron. 2005. *The Changing Role of the Embryo in Evolutionary Thought: Roots of Evo-Devo*. Cambridge University Press.

Angers, Bernard, Emilie Castonguay, and Rachel Massicotte. 2010. Environmentally induced phenotypes and DNA methylation: how to deal with unpredictable conditions until the next generation and after. *Molecular Ecology* 19 (7): 1283–95.

Anstey, Michael L., Stephen M. Rogers, Swidbert R. Ott, Malcolm Burrows, and Stephen J. Simpson. 2009. Serotonin mediates behavioral gregarization underlying swarm formation in desert locusts. *Science* 323 (5914): 627–30.

Athanasiadis, Alekos, Alexander Rich, and Stefan Maas. 2004. Widespread A-to-I RNA editing of alu-containing mRNAs in the human transcriptome. *PLoS Biology* 2 (12): e391.

Badyaev, Alexander V. 2009. Evolutionary significance of phenotypic accommodation in novel environments: an empirical test of the Baldwin effect. *Philosophical Transactions of the Royal Society, Biological Sciences* 364: 1125–41.

2011. Origin of the fittest: link between emergent variation and evolutionary change as a critical question in evolutionary biology. *Proceedings of the Royal Society, Biological Sciences* 278: 1921–9.

Badyaev, Alexander V., and Tobias Uller. 2009. Parental effects in ecology and evolution: mechanisms, processes, and implications. *Philosophical Transactions of the Royal Society, Biological Sciences* 364: 1169–77.

Baetu, Tudor. 2012. Genomic programs as mechanism schemas: a non-reductionist interpretation. *British Journal for the Philosophy of Science*, 63 (3): 649–71.

Barak, Michal, Erez Y. Levanon, Eli Eisenberg, Nurit Paz, Gideon Rechavi, George M. Church, and Ramit Mehr. 2009. Evidence for large diversity in the human transcriptome created by Alu RNA editing. *Nucleic Acids Research* 37 (20): 6905–15.

Baranov, Pavel V., Olga L. Gurvich, Andrew W. Hammer, Raymond F. Gesteland, and John F. Atkins. 2003. Recode 2003. *Nucleic Acids Research* 31 (1): 87–9.

Bass, Brenda. 2001. *RNA Editing: Frontiers in Molecular Biology*. Oxford University Press.

Bateson, P., D. Barker, T. Clutton-Brock, D. Deb, B. D'Udine, R. A. Foley, P. D. Gluckman, K. Godfrey, T. Kirkwood, M. M. Lahr, J. McNamara, N. B. Metcalfe, P. Monaghan, H. G. Spencer, and S. E. Sultan. 2004. Developmental plasticity and human health. *Nature* 430: 419–21.

Bateson, Patrick P. G., and Peter D. Gluckman. 2011. *Plasticity, Robustness, Development and Evolution*. Cambridge University Press.

Bateson, Patrick P. G., and Paul Martin. 1999. *Design for a Life: How Behavior and Personality Develop*. London: Jonathan Cape.

Bateson, William. 1894. *Materials for the Study of Variation*. London: Macmillan.

Beatty, John. 1990. Evolutionary anti-reductionism: historical reflections. *Biology and Philosophy* 5 (2): 199–210.

Bechtel, William. 2006. *Discovering Cell Mechanisms: The Creation of Modern Cell Biology*. Cambridge University Press.

    2007. Reducing psychology while maintaining its autonomy via mechanistic explanation. In *The Matter of the Mind: Philosophical Essays on Psychology, Neuroscience and Reduction*, ed. M. Schouten and H. L. de Jong. Oxford: Blackwell, 172–98. Also at http://mechanism.ucsd.edu.

    2008. *Mental Mechanisms*. New York: Routledge.

Bechtel, William, and Adele Abrahamsen. 2005. Explanation: a mechanistic alternative. *Studies in History and Philosophy of the Biological and Biomedical Sciences* 36: 421–41.

    in press. Thinking dynamically about biological mechanisms: networks of coupled oscillators. *Foundations of Science*.

Bechtel, William, and Richard C. Richardson. 1993. *Discovering Complexity: Decomposition and Localization as Strategies in Scientific Research*. Princeton University Press.

Bentley, David. 2002. The mRNA assembly line: transcription and processing machines in the same factory. *Current Opinion In Cell Biology* 14 (3): 336–42.

Bergstrom, Carl, and Martin Rosvall. 2009. The transmission sense of information. *Biology and Philosophy* 26 (2): 159–76.

Biliya, Shweta, and Lee A. Bulla Jr. 2010. Genomic imprinting: the influence of differential methylation in the two sexes. *Experimental Biology and Medicine* 235 (2): 139–47.

Bissell, Mina J. 1981. The differentiated state of normal and malignant cells or how to define a 'normal' cell in culture. *International Review of Cytology* 70: 27–100.

2003. Tissue specificity: structural cues allow diverse phenotypes from a constant genotype. In *Origination of Organismal Form: Beyond the Gene in Developmental and Evolutionary Biology*, ed. G. B. Müller and S. A. Newman. Cambridge, MA: MIT Press, 103–17.

Blumenthal, Thomas, Donald Evans, Christopher D. Link, Alessandro Guffanti, Daniel Lawson, Jean Thierry-Mieg, Danielle Thierry-Mieg, Wei Lu Chiu, Kyle Duke, Moni Kiraly, and Stuart K. Kim. 2002. A global analysis of Caenorhabditis elegans operons. *Nature* 417 (6891): 851–4.

Bolker, Jessika. 1995. Model systems in developmental biology. *Bioessays* 17: 4515.

Bonasio, Roberto, Shengjiang Tu, and Danny Reinberg. 2010. Molecular signals of epigenetic states. *Science* 330 (6004): 612–16.

Bonduriansky, Russell. 2012. Rethinking heredity, again. *TREE* 27 (6): 330–6.

Bradbury, Jane. 2005. Alternative mRNA splicing: control by combination. *PLoS Biology* 3 (11): e369.

Braendle, Christian, and Thomas Flatt. 2006. A role for genetic accommodation in evolution? *Bioessays* 28: 868–73.

Brandon, Robert N., and Daniel W. McShea. 2010. *Biology's First Law*. University of Chicago Press.

Brentano, Franz. 1874. *Psychologie vom empirischen Standpunkt*. Leipzig: Duncker & Humbolt.

Brigandt, Ingo, and Alan C. Love. 2008. Reductionism in biology. In *The Stanford Encyclopedia of Philosophy (Fall 2008 Edition)*, ed. E. N. Zalta. At http://plato.stanford.edu/archives/fall2008/entries/reduction-biology/.

2010. Evolutionary novelty and the evo-devo synthesis: field notes. *Evolutionary Biology* 37: 93–9.

Buchler, Nicolas E., Ulrich Gerland, and Terence Hwa. 2003. On schemes of combinatorial transcription logic. *Proceedings of the National Academy of Sciences of the United States of America* 100: 5136–41.

Burian, Richard M. 2004. Molecular epigenesis, molecular pleiotropy, and Molecular Gene Definitions. *History and Philosophy of the Life Sciences* 26 (1): 59–80.

Carroll, Sean B. 2005. Evolution at two levels: on genes and form. *PLoS Biology* 3 (7): e245.

Carroll, Sean B., Jennifer K. Grenier, and Scott D. Weatherbee. 2001. *From DNA to Diversity: Molecular Genetics and the Evolution of Animal Design*. Malden, MA: Blackwell.

2005. *From DNA to Diversity: Molecular Genetics and the Evolution of Animal Design*, 2nd edn. Malden, MA: Blackwell.

Caudevilla, C., C. Codony, D. Serra, G. Plasencia, R. Roman, A. Graessmann, G. Asins, M. Bach-Elias, and F. G. Hegardt. 2001. Localization of an exonic splicing enhancer responsible for mammalian natural trans-splicing. *Nucleic Acids Research* 29 (14): 3108–15.

Cawley, Simon, Stefan Bekiranov, Huck H. Ng, Philipp Kapranov, Edward A. Sekinger, Dione Kampa, Antonio Piccolboni, Victor Sementchenko, Jill Cheng, Alan J. Williams, Raymond Wheeler, Brant Wong, Jorg Drenkow, Mark Yamanaka, Sandeep Patel, Shane Brubaker, Hari Tammana, and Thomas R. Gingeras. 2004. Unbiased mapping of transcription factor binding sites along human chromosomes 21 and 22 points to widespread regulation of noncoding RNAs. *Cell* 116: 499–509.

Celottoa, Alicia M., and Brenton R. Graveley. 2001. Alternative splicing of the Drosophila Dscam pre-mRNA is both temporally and spatially regulated. *Genetics* 159: 599–608.

Chadarevian, S. de. 1998. Of worms and programs: *Caenorhabitis Elegans* and the study of development. *Studies in History and Philosophy of the Biological and Biomedical Sciences* 29 (1): 81–105.

Champagne, Francis A., and James P. Curley. 2009a. Epigenetic mechanisms mediating the long-term effects of maternal care on development. *Neuroscience Biobehaviour Review* 33 (4): 593–600.

2009b. The transgenerational influence of maternal care on offspring gene expression and behavior in rodents. In *Maternal Effects in Mammals*, ed. D. Maestripieri and J. M. Mateo. University of Chicago Press, 182–202.

Chapdelaine, Y., and L. Bonen. 1991. The wheat mitochondrial gene for subunit I of the NADH dehydrogenase complex: a trans-splicing model for this gene-in-pieces. *Cell* 65 (3): 465–72.

Chen, Jianjun, Miao Sun, Sanggyu Lee, Guoline Zhou, Janet D. Rowley, and San Ming Wang. 2002. Identifying novel transcripts and novel genes in the human genome by using novel SAGE tags. *Proceedings of the National Academy of Sciences of the United States of America* 99: 12257–62.

Chen, Ling-Ling, and Gordon G. Carmichael. 2010. Decoding the function of nuclear long noncoding RNAs. *Current Opinion in Cell Biology* 22 (3): 357–64.

Cheng, J., P. Kapranov, J. Drenkow, S. Dike, S. Brubaker, S. Patel, J. Long, D. Stern, H. Tammana, G. Helt, et al. 2005. Transcriptional maps of 10 human chromosomes at 5-nucleotide resolution. *Science* 308 (5725): 1149–54.

Chomsky, Noam. 1988. *Language and Problems of Knowledge: The Managua Lectures.* Cambridge, MA: MIT Press.

Chong, Lisa, and L. Bryan Ray. 2002. Whole-istic biology. *Science* 295 (5560): 1661.

Claverie, J. M. 2001. Gene number: what if there are only 30,000 human genes? *Science* 291: 1255–7.

Coelho, Paulo S. R., Anthony C. Bryan, Anuj Kumar, Gerald S. Shadel, and Michael Snyder. 2002. A novel mitochondrial protein, TAR1p, is encoded on the antisense strand of the nuclear 25S rDNA. *Genes and Development* 16: 2755–60.

Collins, Lesley J., Barbara Schönfeld, and Xiaowei Sylvia Chen. 2011. The epigenetics of non-coding RNA. In *Handbook of Epigenetics: The New Molecular and Medical Genetics*, ed. T. Tollefsbol. Amsterdam: Academic Press, 49–61.

Communi, Didier, Nathalie Suarez-Huerta, Danielle Dussossoy, Pierre Savi, and Jean-Marie Boeynaems. 2001. Cotranscription and intergenic splicing of human P2Y(11) SSF1 genes. *Journal of Biological Chemistry* 276 (19): 16561–6.

Conklin, Edwin G. 1908. The mechanism of heredity. *Science* 27: 89–99.

Cope, Edwin Brinker. 1887. *Origin of the Fittest: Essays in Evolution*. New York: D. Appleton.

Craig, Lindsay R. 2010. The so-called extended synthesis and population genetics. *Biological Theory* 5 (2): 117–23.

Craver, Carl F. 2007. *Explaining the Brain: Mechanisms and the Mosaic Unity of Neuroscience*. Oxford: Clarendon Press.

Craver, Carl F., and William Bechtel. 2007. Top-down causation without top-down causes. *Biology and Philosophy* 22 (4), 547–63.

Crick, Francis H. C. 1958. On protein synthesis. *Symposia of the Society for Experimental Biology* 12: 138–63.

    1970. Central Dogma of molecular biology. *Nature* 227: 561–3.

Cummins, Robert. 1975. Functional analysis. *Journal of Philosophy* 72: 741–65.

Darden, Lindley, and Nandy Maull. 1977. Interfield theories. *Philosophy of Science* 44 (1): 43–64.

Darwin, Charles. 1964 [1859]. *On the Origin of Species: A Facsimile of the First Edition*. Cambridge, MA: Harvard University Press.

Davidson, Eric H. 2001. *Genomic Regulatory Systems: Development and Evolution*. San Diego: Academic Press.

Davidson, Eric H., and Douglas H. Erwin. 2006. Gene regulatory networks and the evolution of animal body plans. *Science* 311: 796–800.

Davidson, Eric H., and Michael Levine. 2005. Gene regulatory networks. *PNAS* 102 (14): 4935.

Davidson, Nicholas O. 2002. The challenge of target sequence specificity in C→U RNA editing. *Journal of Clinical Investigation* 109 (3): 291–4.

Dawkins, Richard. 1976. *The Selfish Gene*. Oxford University Press.

Depew, David J., and Bruce H. Weber. 1995. *Darwinism Evolving: Systems Dynamics and the Genealogy of Natural Selection*. Cambridge, MA: Bradford Books/MIT Press.

Devlin, Bernie, Michael Daniels, and Katherine Roeder. 1997. The heritability of IQ. *Nature* 388 (6641): 468–71.

Dietrich, Michael R. 2000. From hopeful monsters to homeotic effects: Richard Goldschmidt's integration of development, evolution and genetics. *American Zoologist* 40: 738–47.

Dillon, Niall. 2003. Positions, please … *Nature* 425 (6957): 457.

Djebali, S., C. A. Davis, A. Merkel, A. Dobin, T. Lassmann, A. Mortazavi, A. Tanzer, J. Lagarde, W. Lin, F. Schlesinger, et al. 2012. Landscape of transcription in human cells. *Nature* 489 (7414): 101–8.

Djebali, Sarah, Julien Lagarde, Philipp Kapranov, Vincent Lacroix, Christelle Borel, Jonathan M. Mudge, Cedric Howald, Sylvain Foissac, Catherine Ucla, Jacqueline Chrast, et al. 2012. Evidence for transcript networks composed of chimeric RNAs in human cells. *PLoS ONE* 7 (1): e28213. doi:10.1371/journal.pone.0028213.

Dobzhansky, Theodosius. 1973. Nothing in biology makes sense except in the light of evolution. *American Biology Teacher* 35: 125–9.

Donnellan, Keith S. 1966. Reference and definite descriptions. *The Philosophical Review* 75 (3): 281–304.

Doolittle, W. Ford, and Carmen Sapienza. 1980. Selfish genes, the phenotype paradigm and genome evolution. *Nature* 284 (5757): 601–3.

Dorn, Rainer, Gunter Reuter, and Andrea Loewendorf. 2001. Transgene analysis proves mRNA trans-splicing at the complex mod(mdg4) locus in Drosophila. *Proceedings of the National Academy of Sciences of the United States of America* 98 (17): 9724–9.

Downes, Stephen. 1992. The importance of models in theorizing: a deflationary semantic view. In *Proceedings of the Philosophy of Science Association, Vol. 1*, ed. D. Hull, M. Forbes and K. Okruhlik. East Lansing, MI: Philosophy of Science Association, 142–53.

Economist. 1999. Drowning in data. *Economist*, 26 June, 97–8.

Eddy, Sean R. 2001. Non-coding RNA genes and the modern RNA world. *Nature Review Genetics* 2 (12): 919–29.

2002. Computational genomics of noncoding RNA genes. *Cell* 109 (2): 137–40.

Eisenberg, David, Edward M. Marcotte, Andrew D. McLachlan, and Matteo Pellegrini. 2006. Bioinformatic challenges for the next decade(s). *Philosophical Transactions of the Royal Society, Biological Sciences* 361 (1467): 525.

ENCODE Project Consortium. 2007. Identification and analysis of functional elements in 1% of the human genome by the ENCODE pilot project. *Nature* 447: 799–816.

Falk, Raphael. 1984. The gene in search of an identity. *Human Genetics* 68: 195–204.

1986. What is a gene? *Studies in the History and Philosophy of Science* 17: 133–73.

2000. The gene: a concept in tension. In *The Concept of the Gene in Development and Evolution*, ed. P. Beurton, R. Falk and H.-J. Rheinberger. Cambridge University Press, 317–48.

2004. Long live the genome! So should the gene. *History and Philosophy of the Life Sciences* 26 (1): 105–21.

2007. Genetic analysis. In *Philosophy of Biology*, ed. M. Matthen and C. Stephens. Oxford: Elsevier, 249–308.

2009. *Genetic Analysis: A History of Genetic Thinking*. Cambridge University Press.

Finta, Csaba, and Peter G. Zaphiropoulos. 2000a. The human CYP2C locus: a prototype for intergenic and exon repetition splicing events. *Genomics* 63 (3): 433–8.

2000b. The human cytochrome P450 3A locus: gene evolution by capture of downstream exons. *Gene* 260 (1–2): 13–23.

2001. A statistical view of genome transcription. *Journal of Molecular Evolution* 53: 160–2.

2002. Intergenic mRNA molecules resulting from trans-splicing. *Journal of Biological Chemistry* 277 (8): 5882–90.

Fisher, R. A. 1918. The correlation between relatives on the supposition of Mendelian inheritance. *Transactions of the Royal Society of Edinburgh* 52: 399–433.

Fodor, Jerry A. 1974. Special sciences. *Synthese* 28: 77–115.

Fogle, Thomas. 2000. The dissolution of protein coding genes in molecular biology. In *The Concept of the Gene in Development and Evolution*, ed. P. Beurton, R. Falk and H.-J. Rheinberger. Cambridge University Press, 3–25.

Fontana, Walter, and Leo W. Buss. 1994. 'The arrival of the fittest': toward a theory of biological organization. *Bulletin of Mathematical Biology* 56 (1): 164.

Ford, Donald H., and Richard M. Lerner. 1992. *Developmental Systems Theory: An Integrative Approach*. Newbury Park, CA: Sage.

Fox, Charles W., Monica S. Thakar, and Timothy A. Mousseau. 1997. Egg size plasticity in a seed beetle: an adaptive maternal effect. *The American Naturalist* 149 (1): 149–63.

Fox, Charles W., Kim J. Waddell, and Timothy A. Mousseau. 1995. Parental host plant affects offspring life histories in a seed beetle. *Ecology* 76 (2): 402–11.

Frenkel-Morgenstern, M., V. Lacroix, I. Ezkurdia, Y. Levin, A. Gabashvili, J. Prilusky, A. Del Pozo, M. Tress, R. Johnson, R. Guigo, et al. 2012. Chimeras taking shape: potential functions of proteins encoded by chimeric RNA transcripts. *Genome Research* 22: 1231–42.

Frigg, Roman. 2010. Models and fiction. *Synthese* 17 (2): 251–68.

Frigg, Roman, and Stephan Hartmann. 2009. *Models in Science*. At http://plato. stanford.edu/archives/sum2009/entries/models-science/.

Fuller, Trevor, Sahotra Sarkar, and David Crews. 2005. The use of norms of reaction to analyze genotypic and environmental influences on behavior in mice and rats. *Neuroscience and Biobehavioral Reviews* 29: 445–56.

Gagen, Michael J., and John S. Mattick. 2004. Inherent size constraints on prokaryote gene networks due to 'accelerating' growth. *Theory in Biosciences* 123(4): 381–411.

Garden, George. 1691. A discourse concerning the modern theory of generation. *Philosophical Transactions of the Royal Society* 16 (1686–1692): 474–83.

Gehring, Walter J. 1985. The homeo box: a key to the understanding of development? *Cell* 40: 3–5.

Gerstein, Mark B., Can Bruce, Joel S. Rozowsky, Deyou Zheng, Jiang Du, Jan O. Korbel, Olof Emanuelsson, Zhengdong D. Zhang, Sherman Weissman, and Michael Snyder. 2007. What is a gene, post-ENCODE? History and updated definition. *Genome Research* 17 (6): 669–81.

Gibbs, W. Wayt. 2003. The unseen genome: gems among the junk. *Scientific American* 289 (5): 46–53.

Giere, Ronald. 1988. *Explaining Science: A Cognitive Approach*. University of Chicago Press.

Gilbert, Scott F. 2000. *Developmental Biology*. Sunderland, MA: Sinauer Associates.

    2001. Ecological developmental biology: developmental biology meets the real world. *Developmental Biology* 233: 1–22.

    2003. The reactive genome. In *Origination of Organismal Form: Beyond the Gene in Developmental and Evolutionary Biology*, ed. G. B. Müller and S. A. Newman. Cambridge, MA: MIT Press, 87–101.

Gilbert, Scott F., and David Epel. 2009. *Ecological Developmental Biology: Integrating Epigenetics, Medicine, and Evolution*. Sunderland, MA: Sinauer Associates.

Gilbert, Scott F., and Sahotra Sarkar. 2000. Embracing complexity: organicism for the twenty-first century. *Developmental Dynamics* 219: 1–9.

Gissis, Snait B., and Eva Jablonka (eds.). 2011. *Transformations of Lamarckism: From Subtle Fluids to Molecular Biology*. Cambridge, MA: MIT Press.

Gluckman, Peter D., and Mark A. Hanson. 2005a. *The Fetal Matrix: Evolution, Development and Disease*. Cambridge University Press.

    (eds.). 2005b. *Developmental Origin of Health and Disease*. Cambridge University Press.

    2006. *Mismatch: Why Our World No Longer Fits Our Bodies*. Oxford University Press.

Gluckman, Peter D., Mark A. Hanson, Patrick Bateson, Alan S. Beedle, Catherine M. Law, Zulfiqar A. Bhutta, Konstantin V. Anokhin, Pierre Bougnères, Giriraj

Ratan Chandak, Partha Dasgupta, George Davey Smith, Peter T. Ellison, Terrence E. Forrester, Scott F. Gilbert, Eva Jablonka, Hillard Kaplan, Andrew M. Prentice, Stephen J. Simpson, Ricardo Uauy, and Mary Jane West-Eberhard. 2009. Towards a new developmental synthesis: adaptive developmental plasticity and human disease. *Lancet* 373 (9675): 1654–7.

Gluckman, Peter D., Mark A. Hanson, and Alan S. Beedle. 2007. Early life events and their consequences for later disease: a life history and evolutionary perspective. *American Journal of Human Biology* 19: 1–19.

Gluckman, Peter D., Mark A. Hanson, and Hamish G. Spencer. 2005. Predictive adaptive responses and human evolution. *Trends in Ecology and Evolution* 20 (10): 527–33.

Godfrey, K. M., A. Sheppard, P. D. Gluckman, K. A. Lillycrop, G. C. Burdge, C. McLean, J. Rodford, J. L. Slater-Jefferies, E. Garratt, S. R. Crozier, B. S. Emerald, C. R. Gale, H. M. Inskip, C. Cooper, and M. A. Hanson. 2011. Epigenetic gene promoter methylation at birth is associated with child's later adiposity. *Diabetes* 30 (5): 1528–34.

Godfrey-Smith, Peter. 1989. Misinformation. *Canadian Journal of Philosophy* 19 (5): 533–50.

1999. Genes and codes: lessons from the philosophy of mind? In *Biology Meets Psychology: Constraints, Conjectures, Connections*, ed. V. G. Hardcastle. Cambridge, MA: MIT Press, 305–31.

2000a. On the theoretical role of 'genetic coding'. *Philosophy of Science* 67 (1): 26–44.

2000b. 'Explanatory symmetries, preformation, and developmental systems theory.' *Philosophy of Science* 67 (Supplementary Proceedings of the 1998 Biennial Meetings of the Philosophy of Science Association. Part II: Symposia Papers): S322–S331.

2001. On the status and explanatory structure of developmental systems theory. In *Cycles of Contingency: Developmental Systems and Evolution*, ed. S. Oyama, P. E. Griffiths and R. D. Gray. Cambridge, MA: MIT Press, 283–97.

2006. The strategy of model-based science. *Biology and Philosophy* 21: 725–40.

2011. Senders, receivers, and genetic information: comments on Bergstrom and Rosvall. *Biology and Philosophy* 26 (2): 177–81.

Goodwin, Brian C. 1994. *How the Leopard Changed Its Spots: The Evolution of Complexity.* New York: Charles Scribner & Sons.

Gott, Jonathan M., and Ronald B. Emeson. 2000. Functions and mechanisms of RNA editing. *Annual Review of Genetics* 34: 499–531.

Gottlieb, Gilbert. 1992. *Individual Development and Evolution.* Oxford University Press.

1995. Some conceptual deficiencies in 'developmental' behavior genetics. *Human Development* 38: 131–41.

1997. *Synthesizing Nature–Nurture: Prenatal Roots of Instinctive Behavior*. Hillsdale, NJ: Lawrence Erlbaum Associates.

2001. A developmental psychobiological systems view: early formulation and current status. In *Cycles of Contingency: Developmental Systems and Evolution*, ed. S. Oyama, P. E. Griffiths and R. D. Gray. Cambridge, MA: MIT Press, 41–54.

2003. On making behavioral genetics truly developmental. *Human Development* 46 (6): 337–55.

Gould, Stephen Jay. 1977. *Ontogeny and Phylogeny*. Cambridge, MA: Harvard University Press.

Graveley, Brenton R. 2005. Mutually exclusive splicing of the insect Dscam pre-mRNA directed by competing intronic RNA secondary structures. *Cell* 123: 65–73.

Gray, M. W. 2003. Diversity and evolution of mitochondrial RNA editing systems. *IUBMB Life* 55 (4–5): 227–33.

Gray, Russell D. 1992. Death of the gene: developmental systems strike back. In *Trees of Life: Essays in the Philosophy of Biology*, ed. Paul E. Griffiths. Dordrecht: Kluwer, 165–210.

Greenspan, Ralph J. 2001. The flexible genome. *Nature Reviews Genetics* 2: 383–7.

Grice, H. P. 1957. Meaning. *Philosophical Review* 66: 377–88.

Griffiths, Paul E. 2001. Genetic information: a metaphor in search of a theory. *Philosophy of Science* 68 (3): 394–412.

2003. Beyond the Baldwin effect: James Mark Baldwin's 'social heredity', epigenetic inheritance and niche-construction. In *Evolution and Learning: The Baldwin Effect Reconsidered*, ed. B. H. Weber and D. J. Depew. Cambridge, MA: MIT Press, 193–215.

2004. Instinct in the '50s: The British reception of Konrad Lorenz's theory of instinctive behaviour. *Biology and Philosophy* 19 (4): 609–31.

2009. In what sense does 'nothing in biology make sense except in the light of evolution'? *Acta Biotheoretica* 57 (1–2): 11–32.

Griffiths, Paul E., and Russell D. Gray. 1994. Developmental systems and evolutionary explanation. *Journal of Philosophy* 91 (6): 277–304.

1997. Replicator II: judgment day. *Biology and Philosophy* 12 (4): 471–92.

2005. Three ways to misunderstand developmental systems theory. *Biology and Philosophy* 20 (2): 417–25.

Griffiths, Paul E., and Rob D. Knight. 1998. What is the developmentalist challenge? *Philosophy of Science* 65 (2): 253–8.

Griffiths, Paul E., and Karola Stotz. 2006. Genes in the postgenomic era. *Theoretical Medicine and Bioethics* 27 (6): 499–521.

2007. Gene. In *Cambridge Companion to the Philosophy of Biology*, ed. D. Hull and M. Ruse. Cambridge University Press, 85–102.

Griffiths, Paul E., and James G. Tabery. 2008. Behavioral genetics and development. *New Ideas in Psychology* 26 (3): 332–52.

Haack, Susan. 1987. Surprising noises: Rorty and Hesse on metaphor. *Proceedings of the Aristotelian Society* 88: 293–301.

Hacking, Ian. 1983. *Representing and Intervening: Introductory Topics in the Philosophy of Natural Science*. Cambridge University Press.

Haig, David. 2000. The kinship theory of genomic imprinting. *Annual Review of Ecology and Systematics* 31: 9–32.

   2004. The (dual) origin of epigenetics. *Cold Spring Harbor Symposia on Quantitative Biology* 69: 67–70.

Haldane, J. B. S., and Helen Spurway. 1954. A statistical analysis of communication in 'Apis mellifera' and a comparison with communication in other animals. *Insectes Sociaux* 1 (3): 247–83.

Hall, Brian K. 1992. *Evolutionary Developmental Biology*. New York: Chapman & Hall.

   1998. *Evolutionary Developmental Biology*. 2nd edn. New York: Chapman & Hall.

   2011. A brief history of the term and concept of epigenetics. In *Epigenetics: Linking Genotype and Phenotype in Development and Evolution*, ed. B. Hallgrimsson and B. K. Hall. Berkeley: University of California Press, 9–13.

Hallgrimsson, Benedict, and Brian K. Hall. 2011. Introduction. In *Epigenetics: Linking Genotype and Phenotype in Develoment and Evolution*, ed. B. Hallgrimsson and B. K. Hall. Berkeley: University of California Press, 1–5.

Hamer, Dean. 2002. Rethinking behavior genetics. *Science* 298: 71–2.

Hardison, Ross C. 2000. Conserved non-coding sequences are reliable guides to regulatory elements. *Trends in Genetics* 16 (9): 369–72.

Harrison, Paul M., Anuj Kumar, Ning Lang, Michael Snyder, and Mark Gerstein. 2002. A question of size: the eukaryotic proteome and the problems in defining it. *Nucleic Acids Research* 30 (5): 1083–90.

Häsler, J., T. Samuelsson, and K. Strub. 2007. Useful 'junk': Alu RNAs in the human transcriptome. *Cellular and Molecular Life Sciences* 64 (14): 1793–800.

Hauber, M. E., P. W. Sherman, and D. Paprika. 2000. Self-referent phenotype matching in a brood parasite: the armpit effect in brown-headed cowbirds (*Molothrus ater*). *Animal Cognition* 3 (2): 113–17.

Helden, Jacques van, Avi Naim, Renato Mancuso, Matthew Eldridge, Lorenz Wernisch, David Gilbert, and Shoshana J. Wodak. 2000. Representing and analysing molecular and cellular function using the computer. *Journal of Biological Chemistry* 381: 921–35.

Herring, Susan W. 1993. Formation of the vertebrate face: epigenetic and functional influences. *American Zoologist* 33 (4): 472–83.

Hesse, Mary B. 1966. *Models and Analogies in Science*. University of Notre Dame Press.

1988. The cognitive claims of metaphor. *Journal of Speculative Philosophy* 2 (1): 1–16.

Hirotsune, Shinji, Noriyuki Yoshida, Amy Chen, Lisa Garrett, Fumihiro Sugiyama, Satoru Takahashi, Ken-ichi Yagami, Anthony Wynshaw-Boris, and Atsushi Yoshiki. 2003. An expressed pseudogene regulates the messenger-RNA stability of its homologous coding gene. *Nature* 423 (6935): 91–6.

Holliday, Robin. 1987. The inheritance of epigenetic defects. *Science* 238 (4824): 163–70.

1994. Introduction: epigenetics, an overview. *Developmental Genetics* 15: 453–7.

2006. Epigenetics: a historical overview. *Epigenetics* 1 (2): 76–80.

Hood, Kathryn E., Carolyn Tucker Halpern, Gary Greenberg, and Richard M. Lerner (eds.). 2010. *Handbook of Developmental Science, Behavior and Genetics*. Oxford: Wiley-Blackwell.

Hull, David. 1972. Reduction in genetics – biology or philosophy? *Philosophy of Science* 39 (4): 491–9.

1974. *Philosophy of Biological Science*. Englewood Cliffs, NJ: Prentice-Hall.

1975. Informal aspects of theory reduction. In *Boston Studies in the Philosophy of Science; Proceedings of the Biennial Meeting of the Philosophy of Science Association, 1974*, ed. R. S. Cohen and A. Michalos. Dordrecht: Reidel, 653–70.

Hundley, Heather A., and Brenda L. Bass. 2010. ADAR editing in double-stranded UTRs and other noncoding RNA sequences. *Trends in Biochemical Sciences* 35 (7): 377–83.

Hüttenhofer, A., M. Kiefmann, S. Meier-Ewert, J. O'Brien, H. Lehrach, J. P. Bachellerie, and J. Brosius. 2001. RNomics: an experimental approach that identifies 201 candidates for novel, small, non-messenger RNAs in mouse. *EMBO Journal* 20 (11): 2943–53.

Huxley, Julian. 1972 [1932]. *Problems of Relative Growth*. New York: Dover.

Istrail, Sorin, and Eric H. Davidson. 2005. Logic functions of the genomic cis-regulatory code. *Proceedings of the National Academy of Sciences* 102 (14): 4954–9.

Jablonka, Eva. 2001. The systems of inheritance. In *Cycles of Contingency: Developmental Systems and Evolution*, ed. S. Oyama, P. E. Griffiths and R. D. Gray. Cambridge, MA: MIT Press, 99–116.

2002. Information interpretation, inheritance, and sharing. *Philosophy of Science* 69 (4): 578–605.

Jablonka, Eva, and Eytan Avital. 2001. *Animal Traditions : Behavioural Inheritance in Evolution*. Cambridge University Press.

Jablonka, Eva, and Marion J. Lamb. 1995. *Epigenetic Inheritance and Evolution: The Lamarckian Dimension*. Oxford University Press.

2005. *Evolution in Four Dimensions: Genetic, Epigenetic, Behavioral, and Symbolic Variation in the History of Life*. Cambridge, MA: MIT Press.

Jablonka, Eva, and Gal Raz. 2009. Transgenerational epigenetic inheritance: prevalence, mechanisms, and implications for the study of heredity and evolution. *Quarterly Review of Biology* 84 (2): 131–76.

Jacob, François, and Jacques Monod. 1961. Genetic regulatory mechanisms in the synthesis of proteins. *Journal of Molecular Biology* 3: 318–56.

Jenuwein, Thomas, and C. David Allis. 2001. Translating the histone code. *Science* 293: 1074–80.

Johannsen, William. 1911. The genotype conception of heredity. *American Naturalist* 45 (531): 129–59.

Johnson, J. M., S. Edwards, D. D. Shoemaker, and E. E. Schadt. 2005. Dark matter in the genome: evidence of widespread transcription detected by microarray tiling experiments. *Trends in Genetics* 21: 93–102.

Johnston, Timothy D. 1987. The persistence of dichotomies in the study of behavioural development. *Developmental Review* 7: 149–82.

Juarrero, Alicia. 1998. Causality as constraints. In *Evolving Systems*, ed. G. Van de Vijver, S. Salthe, and M. Delpos. Dordrecht: Kluwer.

Kampa, Dione, Jill Cheng, Philipp Kapranov, Mark Yamanaka, Shane Brubaker, Simon Cawley, Jorg Drenkow, Antonio Piccolboni, Stefan Bekiranov, Gregg Helt, Hari Tammana, and Thomas R. Gingeras. 2004. Novel RNAs identified from an in-depth analysis of the transcriptome of human chromosomes 21 and 22. *Genome Research* 14: 331–42.

Kaplan, Jonathan Michael. 2000. *The Limits and Lies of Human Genetic Research.* Oxford: Routledge.

Kapranov, Phillip, Jorg Drenkow, Jill Cheng, Jeffrey Long, Gregg Helt, Sujit Dike, and Thomas R. Gingeras. 2005. Examples of the complex architecture of the human transcriptome revealed by RACE and high-density tiling arrays. *Genome Research* 15: 987–97.

Kauffman, Stuart A. 1993. *The Origins of Order: Self-Organisation and Selection in Evolution.* New York: Oxford University Press.

Kay, Lily E. 2000. *Who Wrote the Book of Life: A History of the Genetic Code.* Palo Alto, CA: Stanford University Press.

Kendler, Kenneth S. 2005. 'A gene for . . .': the nature of gene action in psychiatric disorders. *American Journal of Psychiatry* 162 (7): 1243–52.

Kim, Jaegwon. 1993. *Supervenience and Mind: Selected Philosophical Essays.* Cambridge University Press.

Kirschner, Marc W. and John C. Gerhart. 1998. Evolvability. *Proceedings of the National Academy of Sciences of the United States of America* 95, 8420–7.

2005. *The Plausibility of Life: Resolving Darwin's Dilemma.* London: Yale University Press.

2010. Facilitated variation. In *Evolution: The Extended Synthesis*, ed. G. B. Mueller and M. Pigliucci. Cambridge, MA: MIT Press.

Kirschner, Marc W., John C. Gerhart, and Tim Mitchison. 2000. Molecular 'vitalism'. *Cell* 100: 79–88.

Kitcher, Philip. 1982. Genes. *British Journal of Philosophy of Science* 33: 337–59.

1984. 1953 and all that: a tale of two sciences. *Philosophical Review* 93: 335–73.

2001. Battling the undead: how (and how not) to resist genetic determinism. In *Thinking about Evolution: Historical, Philosophical and Political Perspectives (Festschrift for Richard Lewontin)*, ed. R. Singh, K. Krimbas, D. Paul and J. Beatty. Cambridge University Press, 369–414.

Kohler, Robert E. 1994. *Lords of the Fly. Drosophila Genetics and the Experimental Life.* University of Chicago Press.

Kroh, Evan M., Rachael K. Parkin, Patrick S. Mitchell, and Muneesh Tewari. 2010. Analysis of circulating microRNA biomarkers in plasma and serum using quantitative reverse transcription-PCR (qRT-PCR). *Methods* 50 (4): 298–301.

Kuhn, Thomas. 1962. *The Structure of Scientific Revolutions.* University of Chicago Press.

Lacey, Elizabeth R. 1998. What is an adaptive environmentally induced parental effect? In *Maternal Effects as Adaptations*, ed. R. A. Mousseau and C. W. Fox. New York: Oxford University Press, 54–66.

Ladd, Andrea N., and Thomas A. Cooper. 2002. Finding signals that regulate alternative splicing in the post-genomic era. *Genome Biology* 3 (11): 1–6.

Laland, Kevin N., F. John Odling-Smee, and Scott F. Gilbert. 2008. EvoDevo and niche construction: building bridges. *Journal of Experimental Zoology, Part B: Molecular and Developmental Evolution* 310: 549–66.

Laland, Kevin N., Kim Sterelny, John Odling-Smee, W. Hoppitt, and Tobias Uller. 2011. Cause and effect in biology revisited: is Mayr's proximate-ultimate dichotomy still useful? *Science* 334 (6062): 1512–16.

Lamm, Ehud, and Eva Jablonka. 2008. The nurture of nature: hereditary plasticity in evolution. *Philosophical Psychology* 21 (3): 305–19.

Lande, R., and T. Price. 1989. Genetic correlations and maternal effect coefficients obtained from offspring-parent regression. *Genetics* 122 (4): 915–22.

Lander, E. S., L. M. Linton, B. Birren, C. Nusbaum, M. C. Zody, J. Baldwin, K. Devon, K. Dewar, M. Doyle, W. FitzHugh, et al. 2001. Initial sequencing and analysis of the human genome. *Nature* 409: 860–921.

Lander, Eric S., and Robert A. Weinberg. 2000. Journey to the centre of biology. *Science* 287 (5459): 1777–82.

Lederberg, Joshua. 1956a. Comments on gene–enzyme relationship. In *Enzymes: Units of Biological Structure and Function*, ed. E. H. Gaebler. New York: Academic Press, 161–9.

1956b. Genetic transduction. *American Scientist* 44 (3): 264–80.

1996. What the double helix (1953) has meant for basic biomedical science: a personal commentary. In *The Philosophy and History of Molecular Biology: New Perspectives*, ed. S. Sarkar. Dordrecht: Kluwer, 15–26.

Lehrman, Daniel S. 1970. Semantic and conceptual issues in the nature–nurture problem. In *Development and Evolution of Behaviour*, ed. D. S. Lehrman. San Francisco: W. H. Freeman, 17–52.

Leipzig, Jeremy, Pavel Pevzner, and Steffen Heber. 2004. The alternative splicing gallery (ASG): bridging the gap between genome and transcriptome. *Nucleic Acids Research* 32 (13): 3977–83.

Lettice, Laura A., Taizo Horikoshi, Simon J. H. Heaney, Marijke J. van Baren, Herma C. van der Linde, Guido J. Breedveld, Marijke Joosse, Nurten Akarsu, Ben A. Oostra, Naoto Endo, Minoru Shibata, Mikio Suzuki, Eiichi Takahashi, Toshikatsu Shinka, Yutaka Nakahori, Dai Ayusawa, Kazuhiko Nakabayashi, Stephen W. Scherer, Peter Heutink, Robert E. Hill, and Sumihare Noji. 2002. Disruption of a long-range cis-acting regulator for Shh causes preaxial polydactyly. *Proceedings of the National Academy of Sciences* 99 (11): 7548–53.

Levit, Georgy S., Uwe Hossfeld, and Lennart Olsson. 2006. From the 'Modern Synthesis' to cybernetics: Ivan Ivanovich Schmalhausen (1884–1963) and his research program for a synthesis of evolutionary and developmental biology. *Journal of Experimental Zoology, Part B: Molecular and Developmental Evolution* 306: 89–106.

Levy, Arnon. 2011. Information in biology: a fictionalist account. *Nous* 45 (4): 640–57.

Lewis, David K. 1966. An argument for the identity theory. *Journal of Philosophy* 63 (1): 17–25.

Lewontin, Richard C. 1982. Organism and environment. In *Learning, Development, Culture*, ed. H. Plotkin. New York: John Wiley, 151–70.

1983a. Gene, organism and environment. In *Evolution: From Molecules to Man*, ed. D. S. Bendall. Cambridge University Press, 273–85.

1983b. The organism as the subject and object of evolution. *Scientia* 118: 65–82.

1996. Evolution as engineering. In *Integrative Approaches to Molecular Biology*, ed. J. Collado-Vides, B. Magasanik and T. F. Smith. Cambridge, MA: MIT Press, 1–10.

Lipovich, L., R. Johnson, and C. Y. Lin. 2010. MacroRNA underdogs in a microRNA world: evolutionary, regulatory, and biomedical significance of mammalian long non-protein-coding RNA. *Biochimica et Biophysica Acta* 1799 (9): 597–615.

Liu, Yilei, and David J. Elliott. 2010. Coupling genetics and post-genomic approaches to decipher the cellular splicing code at a systems-wide level. *Biochemistry Society Transactions* 38: 237–41.

Lloyd, Elizabeth A. 1988. *The Structure and Confirmation of Evolutionary Theory*. Westport, CT: Greenwood Press.

Lorenz, Konrad Z. 1965. *Evolution and the Modification of Behaviour*. University of Chicago Press.

Love, Alan C. 2001. Evolutionary morphology, innovation, and the synthesis of evolution and development. *Biology and Philosophy* 18 (2): 309–45.

2008. Explaining evolutionary innovation and novelty: criteria of adequacy and multidisciplinary prerequisites. *Philosophy of Science* 75: 874–86.

McEachern, Lori A., and Vett Loyd. 2011. The epigenetics of genomic imprinting. In *Epigenetics: Linking Genotype and Phenotype in Develoment and Evolution*, ed. B. Hallgrimsson and B. K. Hall. Berkeley: University of California Press, 43–69.

Maestripieri, Dario. 2009. Maternal influences in offspring growth, reproduction, and behavior in primates. In *Maternal Effects in Mammals*, ed. D. Maestripieri and J. M. Mateo. University of Chicago Press, 256–91.

Maestripieri, Dario, and Jill M. Mateo (eds.). 2009. *Maternal Effects in Mammals*. University of Chicago Press.

Maher, Brendan. 2012. The human encyclopedia: making a genome manual. *Nature* 489 (7414): 46–8.

Maienschein, Jane. 2005. Epigenesis and preformationism. In *The Stanford Encyclopedia of Philosophy*, ed. E. N. Zalta. At http://plato.stanford.edu/entries/⟨?PMU ?⟩epigenesis/.

Malek, Olaf, and Volker Knoop. 1998. Trans-splicing group II introns in plant mitochondria: the complete set of cis-arranged homologs in ferns, fern allies, and a hornwort. *RNA* 4 (12): 1599–609.

Manolio, Teri A., Francis S. Collins, Nancy J. Cox, David B. Goldstein, Lucia A. Hindorff, David J. Hunter, Mark I. McCarthy, Erin M. Ramos, Lon R. Cardon, Aravinda Chakravarti, Judy H. Cho, Alan E. Guttmacher, Augustine Kong, Leonid Kruglyak, Elaine Mardis, Charles N. Rotimi, Montgomery Slatkin, David Valle, Alice S. Whittemore, Michael Boehnke, Andrew G. Clark, Evan E. Eichler, Greg Gibson, Jonathan L. Haines, Trudy F. C. Mackay, Steven A. McCarroll, and Peter M. Visscher. 2009. Finding the missing heritability of complex diseases. *Nature* 461 (7265): 747–53.

Matthen, Mohan, and André Ariew. 2002. Two ways of thinking about fitness and natural selection. *Journal of Philosophy* 992: 55–83.

Mattick, John S. 2001. Non-coding RNAs: the architects of eukaryotic complexity. *EMBO Reports* 2 (11): 986–91.

2003. Challenging the dogma: the hidden layer of non-protein-coding RNAs in complex organisms. *Bioessays* 25 (10): 930–9.

2004. RNA regulation: a new genetics? *Nature Reviews Genetics* 5 (4): 316–23.

Maynard Smith, John. 2000. The concept of information in biology. *Philosophy of Science* 67 (2): 177–94.

Maynard Smith, J., R. Burian, S. Kauffman, P. Alberch, J. Campbell, B. Goodwin, R. Lande, D. Raup, and L. Wolpert. 1985. Developmental constraints and evolution. *Quarterly Review of Biology* 60 (3): 265–87.

Maynard Smith, John, and Eörs Szathmáry. 1995. *The Major Transitions in Evolution.* New York: W. H. Freeman.

Mayr, Ernst. 1961. Cause and effect in biology. *Science* 134 (3489): 1501–6.

1964. The evolution of living systems. *Proceedings of the National Academy of Sciences of the United States of America* 51 (5): 934–41.

1976 [1959]. Typological versus populational thinking. In *Evolution and the Diversity of Life*, ed. E. Mayr. Cambridge, MA: Harvard University Press, 26–9.

Mayr, Ernst, and William B. Provine. 1980. *The Evolutionary Synthesis: Perspectives on the Unification of Biology.* Cambridge, MA: Harvard University Press, 43–69.

Meaney, Michael J. 2001a. Nature, nurture, and the disunity of knowledge. *Annals of the New York Academy of Sciences* 935 (1): 50–61.

2001b. Maternal care, gene expression, and the transmission of individual differences in stress reactivity across generations. *Annual Review of Neuroscience* 24: 1161–92.

2004. The nature of nurture: maternal effect and chromatin modelling. In *Essays in Social Neuroscience*, ed. J. T. Cacioppo and G. G. Berntson. Cambridge, MA: MIT Press.

Meaney, Michael J., and Moshe Szyf. 2005. Environmental programming of stress responses through DNA methylation: life at the interface between a dynamic environment and a fixed genome. *Dialogues in Clinical Neuroscience* 7 (2): 103–23.

Mendel, Gregor. 1866. Versuche über Pflanzenhybriden [Experiments in plant hybridization]. *Verhandlungen des naturforschenden Vereines in Brünn, Bd. IV für das Jahr 1865*, Abhandlungen: 3–47.

Michel, George F., and Celia L. Moore. 1995. *Developmental Psychobiology: An Interdisciplinary Science.* Cambridge, MA: MIT Press.

Millikan, Ruth Garrett. 1984. *Language, Thought and Other Biological Categories.* Cambridge, MA: MIT Press.

Mitchell, Sandra D. 2009. *Unsimple Truths: Science, Complexity, and Policy.* University of Chicago Press.

Moore, Celia L. 1984. Maternal contributions to the development of masculine sexual behavior in laboratory rats. *Developmental Psychobiology* 17: 346–56.

1992. The role of maternal stimulation in the development of sexual behavior and its neural basis. *Annals of the New York Academy of Sciences* 662: 160–77.

Moore, David S. 2001. *The Dependent Gene: The Fallacy of 'Nature vs. Nurture'*. New York: Henry Holt/Times Books.

Morange, Michel. 2000 [1998]. *A History of Molecular Biology*, trans. Matthew Cobb. Cambridge, MA: Harvard University Press.

2002. The relations between genetics and epigenetics: a historical point of view. *Annals of the New York Academy of Sciences* 981: 50–60.

2008. *Life Explained*. New Haven, CT: Yale University Press.

Morgan, Thomas Hunt. 1926. *The Theory of the Gene*. New Haven, CT: Yale University Press.

1934. The relation of genetics to physiology and medicine. Nobel lecture, 4 June. At www.nobelprize.org/nobel_prizes/medicine/laureates/1933/morgan-lecture.pdf.

Morgan, Thomas Hunt, Alfred H. Sturtevant, H. J. Mullen and Calvin B. Bridges. 1915. *The Mechanism of Mendelian Heredity*. New York: Holt.

Morowitz, Harold J. 2003. *The Emergence of Everything: How the World Became Complex*. Oxford University Press.

Morris, Kevin V. 2008. RNA mediated transcriptional gene silencing. In *RNA and the Regulation of Gene Expression: A Hidden Layer of Complexity*, ed. K. V. Morris. Norfolk, UK: Caister Academic Press, 19–28.

(ed.). 2012. *Non-coding RNAs and Epigenetic Regulation of Gene Expression: Drivers of Natural Selection*. Norfolk, UK: Caister Academic Press.

Moss, Lenny. 2003. *What Genes Can't Do*. Cambridge, MA: MIT Press.

Mousseau, Timothy A., and Charles W. Fox (eds.). 1998. *Maternal Effects as Adaptations*. Oxford University Press.

Mueller, Michael, Lennart Martens, and Rolf Apweiler. 2007. Annotating the human proteome: beyond establishing a parts list. *Biochimica et Biophysica Acta (BBA) – Proteins and Proteomics* 1774 (2): 175–91.

Müller, Gerd B. 1990. Developmental mechanisms as the origin of morphological novelty: a side-effect hypothesis. In *Evolutionary Innovations*, ed. M. H. Nitecki. University of Chicago Press, 99–130.

2003. Homology: the evolution of morphological organization. In *Origination of Organismal Form: Beyond the Gene in Developmental and Evolutionary Biology*, ed. G. B. Müller and S. A. Newman. Cambridge, MA: MIT Press, 51–69.

2010. Epigenetic innovation. In *Evolution – The Extended Synthesis*, ed. M. Pigliucci and G. B. Müller. Cambridge, MA: MIT Press, 307–32.

Müller, Gerd B., and Stuart A. Newman. 2003. Origination of organismal form: the forgotten cause in evolutionary theory. In *Origination of Organismal Form*, ed. G. B. Müller and S. A. Newman. Cambridge, MA: MIT Press, 3–10.

Müller, Gerd B., and Günter P. Wagner. 1991. Novelty in evolution: restructuring the concept. *Annual Review of Ecology and Systematics* 22: 229–56.

Muller, Hermann J. 1947. Pilgrim Trust Lecture: the gene. *Proceedings of the Royal Society of London Series B: Biological Sciences* 134 (874): 1–37.

Müller-Wille, Staffan, and Hans-Jörg Rheinberger. 2012. *A Cultural History of Heredity*. University of Chicago Press.

Nagel, Ernest. 1961. *The Structure of Science: Problems in the Logic of Scientific Explanation*. London: Routledge & Kegan Paul.

Nanney, D. L. 1958. Epigenetic control systems. *Proceedings of the National Academy of Sciences of the United States of America* 44: 712.

Nathanielsz, Peter W., and Kent L. Thornburg. 2003. Fetal programming: from gene to functional systems – an overview. *Journal of Physiology* 547: 3–4.

Nature Genetics. 2010. On beyond GWAS. (Editorial). *Nature Genetics* 42 (7): 551.

Newman, Stuart A. 2003. From physics to development: the evolution of morphogenetic mechanisms. In *Origination of Organismal Form: Beyond the Gene in Developmental and Evolutionary Biology*, ed. G. B. Müller and S. A. Newman. Cambridge, MA: MIT Press, 221–39.

2010. Dynamic patterning modules. In *Evolution – The Extended Synthesis*, ed. M. Pigliucci and G. B. Müller. Cambridge, MA: MIT Press, 281–306.

Newman, Stuart A., and Gerd B. Müller. 2000. The epigenetic basis of character origination. *Journal of Experimental Zoology (Molecular and Developmental Evolution)* 288: 304–17.

Nijhout, H. F. 1990. Metaphors and the role of genes in development. *BioEssays* 12 (9): 441–6.

Noble, Denis. 2002. Modeling the heart: from genes to cells to the whole organ. *Science* 295: 1678–82.

2006. *The Music of Life: Biology beyond Genes*. Oxford University Press.

O'Malley, Maureen A., and John Dupre. 2005. Fundamental issues in systems biology. *BioEssays* 27 (12): 1270–6.

O'Malley, Maureen A., and Karola Stotz. 2011. Intervention, integration and translation in obesity research: genetic, developmental and metaorganismal approaches. *Philosophy, Ethics, and Humanities in Medicine* 6 (2): doi:10.1186/1747-5341-6-2.

Odling-Smee, F. John. 1988. Niche-constructing phenotypes. In *The Role of Behavior in Evolution*, ed. H. C. Plotkin. Cambridge, MA: MIT Press, 73–132.

Odling-Smee, F. John, Kevin N. Laland, and Marcus W. Feldman. 2003. *Niche Construction: The Neglected Process in Evolution*. Princeton University Press.

Olby, Robert C. 1974. *The Path to the Double Helix*. Seattle: University of Washington Press.

1985. *The Origins of Mendelism*, 2nd edn. University of Chicago Press.

2009. *Francis Crick: Hunter of Life's Secret*. Cold Spring Harbor Laboratory Press.

Oliver, Brian, and Benoit Leblanc. 2003. How many genes in a genome? *Genome Biology* 5 (1): 204.1–3.

Oliver, Stephen G. 2006. From genomes to systems: the path with yeast. *Philosophical Transactions of the Royal Society, Biological Sciences* 361: 477–82.

Orgel, Leslie E., and Francis H. Crick. 1980. Selfish DNA: the ultimate parasite. *Nature* 284: 604–7.

Oyama, Susan. 1985. *The Ontogeny of Information: Developmental Systems and Evolution*. Cambridge University Press.

2000a. Causal democracy and causal contributions in developmental systems theory. *Philosophy of Science* 67 (s1): S332.

2000b. *Evolution's Eye: A Systems View of the Biology–Culture Divide*. Durham, NC: Duke University Press.

Oyama, Susan, Paul E. Griffiths, and Russell D. Gray (eds.). 2001. *Cycles of Contingency: Developmental Systems and Evolution*. Cambridge, MA: MIT Press.

Pang, Ken C., Stuart Stephen, Pär G. Engström, Khairina Tajul-Arifin, Weisan Chen, Claes Wahlestedt, Boris Lenhard, Yoshihide Hayashizaki, and John S. Mattick. 2005. RNAdb – a comprehensive mammalian noncoding RNA database. *Nucleic Acids Research* 33 (Database issue): D125–30.

Papineau, David. 1987. *Reality and Representation*. Oxford: Blackwell.

Park, E., B. Williams, B. J. Wold, and A. Mortazavi. 2012. RNA editing in the human ENCODE RNA-seq data. *Genome Research* 22 (9): 1626–33.

Paz-Yaacov, N., E. Y. Levanon, E. Nevo, Y. Kinar, A. Harmelin, J. Jacob-Hirsch, N. Amariglio, E. Eisenberg, and G. Rechavi. 2010. Adenosine-to-inosine RNA editing shapes transcriptome diversity in primates. *Proceedings of the National Academy of Sciences of the United States of America* 107 (27): 12174–9.

Pearson, Karl. 1900 [1892]. *The Grammar of Science*, 2nd edn. London: Black.

Pearson, Karl, G. U. Yule, Norman Blanchard, and Alice Lee. 1903. The law of ancestral heredity. *Biometrika* 2 (2): 211–36.

Pennisi, Elizabeth. 2008. Deciphering the genetics of evolution: powerful personalities lock horns over how the genome changes to set the stage for evolution. *Science* 321: 670–3.

2010. Shining a light on genome's 'dark matter'. *Science* 330: 1614.

Pigliucci, Massimo. 2001. *Phenotypic Plasticity: Beyond Nature and Nurture, Syntheses in Ecology and Evolution*. Baltimore, MD: Johns Hopkins University Press.

2007. Do we need an extended evolutionary synthesis? *Evolution* 61 (12): 2743–9.

2009. An extended synthesis for evolutionay biology. *Annals of the New York Academy of Sciences* 1168: 218–28.

Pigliucci, Massimo, and Jonathan Kaplan. 2006. *Making Sense of Evolution: The Conceptual Foundations of Evolutionary Biology*. University of Chicago Press.

Pigliucci, Massimo, and Gert B. Müller. 2010a. *Evolution – The Extended Synthesis*. Cambridge, MA: MIT Press.

2010b. Elements of an extended evolutionary synthesis. In *Evolution: The Extended Synthesis*, ed. M. Pigliucci and G. B. Müller. Cambridge, MA: MIT Press, 3–17.

Plomin, Robert, and Denise Daniels. 1987. Why are children from the same family so different from one another? *Behavioral and Brain Sciences* 10: 1–60.

Plomin, R., J. C. DeFries, and J. Loehlin. 1977. Genotype-environment interaction and correlation in the analysis of human behavior. *Psychological Bulletin* 84: 309–22.

Ponicsan, Steven L., Jennifer F. Kugel, and James A. Goodrich. 2010. Genomic gems: SINE RNAs regulate mRNA production. *Current Opinion in Genetics and Development* 20 (2): 149–55.

Portin, Petter. 1993. The concept of the gene: short history and present status. *The Quarterly Review of Biology* 68 (2): 173–223.

Ptashne, Mark. 2007. On the use of the word 'epigenetic'. *Current Biology* 17: R233–6.

Ptashne, Mark, and Alexander Gann. 2002. *Genes and Signals*. Cold Spring Harbor Laboratory Press.

Putnam, Hilary. 1975. The meaning of 'meaning'. In *Philosophical Papers. Vol. 2: Mind, Language and Reality*. Cambridge University Press.

Rassoulzadegan, Minoo, Valérie Grandjean, Pierre Gounon, Stéphane Vincent, Isabelle Gillot, and François Cuzin. 2006. RNA-mediated non-mendelian inheritance of an epigenetic change in the mouse. *Nature* 441: 469–74.

Rechavi, Oded, Gregory Minevich, and Oliver Hober. 2011. Transgenerational inheritance of an acquired small RNA-based antiviral response in C. elegans. *Cell* 147 (6): 1248–56.

Reik, Wolf, and Jörn Walter. 2001a. Evolution of imprinting mechanisms: the battle of the sexes begins in the zygote. *Nature Genetics* 27: 255–6.

2001b. Genomic imprinting: parental influence on the genome. *Nature Reviews Genetics* 2: 21–32.

Rivier, Christian, Michel Goldschmidt-Clermont, and Jean-David Rochaix. 2001. Identification of an RNA-protein complex involved in chloroplast group II intron trans-splicing in Chlamydomonas reinhardtii. *EMBO Journal* 20 (7): 1765–73.

Robert, Jason S. 2000. Schizophrenia epigenesis? *Theoretical Medicine and Bioethics* 21 (2): 191–215.

2001. Interpreting the homeobox: metaphors of gene action and activation in development and evolution. *Evolution and Development* 3 (4): 287–95.

2004. *Embryology, Epigenesis and Evolution: Taking Development Seriously.* Cambridge University Press.

Robert, Jason S., Brian K. Hall, and Wendy M. Olson. 2001. Bridging the gap between developmental systems theory and evolutionary developmental biology. *Bioessays* 23 (10): 954–62.

Robinson, Victoria L. 2009. Rethinking the Central Dogma: noncoding RNAs are biologically relevant. *Urologic Oncology* 27 (3): 304–6.

Rosenberg, Alex. 1978. The supervenience of biological concepts. *Philosophy of Science* 45: 368–86.

1985. *The Structure of Biological Science.* Cambridge University Press.

2006. *Darwinian Reductionism: Or, How to Stop Worrying and Love Molecular Biology.* University of Chicago Press.

Rowen, Lee, Janet Young, Brian Birditt, Amardeep Kaur, Anup Madan, Dana L. Philipps, Shizhen Qin, Patrick Minx, Richard K. Wilson, Leroy Hood, and Brenton R. Graveley. 2002. Analysis of the human neurexin genes: alternative splicing and the generation of protein diversity. *Genomics* 79 (4): 587–98.

Ruddle, Frank H., Chris T. Amemiya, Janet L. Carr, Chang-Bae Kim, Christina Ledje, Cooduvalli S. Shashikant, and Guenter P. Wagner. 1999. Evolution of chordate Hox gene clusters. In *Molecular Strategies in Biological Evolution*, ed. L. H. Caporale. New York Academy of Sciences, 238–50.

Ruse, Michael E. 1971. Reduction, replacement, and molecular biology. *Dialectica* 25 (1): 39–72.

Sapp, Jan. 1987. *Beyond the Gene: Cytoplasmic Inheritance and the Struggle for Authority in Genetics.* Oxford University Press.

Sarkar, Sahotra. 1996. Biological information: a skeptical look at some central dogmas of molecular biology. In *The Philosophy and History of Molecular Biology: New Perspectives*, ed. S. Sarkar. Dordrecht: Kluwer, 187–231.

1998. *Genetics and Reductionism.* Cambridge University Press.

1999. From the Reaktionsnorm to the adaptive norm: the norm of reaction 1909–1960. *Biology and Philosophy* 14: 235–52.

2005. *Molecular Models of Life: Philosophical Papers on Molecular Biology.* Cambridge, MA: MIT Press.

Schaffner, Kenneth F. 1967. Antireductionism and molecular biology. In *Man and Nature: Philosophical Issues in Biology*, ed. R. Munson. New York: Dell, 44–54.

1969. The Watson-Crick model and reductionism. *British Journal for the Philosophy of Science* 20: 325–48.

1993. *Discovery and Explanation in Biology and Medicine*. University of Chicago Press.

1996. Theory structure and knowledge representation in molecular biology. In *The Philosophy and History of Molecular Biology: New Perspectives*, ed. S. Sarkar. Dordrecht: Kluwer, 27–46.

1998. Genes, behavior and developmental emergentism: one process, indivisible? *Philosophy of Science* 65 (2): 209–52.

2006a. Behaving: its nature and nurture, part 1. In *Wrestling with Behavioral Genetics: Implications for Understanding Selves and Society*, ed. E. Parens, A. Chapman and N. Press. Baltimore, MD: Johns Hopkins University Press, 3–39.

2006b. Behaving: its nature and nurture, part 2. In *Wrestling with Behavioral Genetics: Implications for Understanding Selves and Society*, ed. E. Parens, A. Chapman and N. Press. Baltimore, MD: Johns Hopkins University Press, 40–73.

Forthcoming. *Behaving: What's Genetic, What's Not, and Why Should We Care?* New York: Oxford University Press.

Schlichting, Carl D., and Massimo Pigliucci. 1998. *Phenotypic Evolution: A Reaction Norm Perspective*. Sunderland, MA: Sinauer.

Schlitt, Thomas, and Alvis Brazma. 2006. Modelling in molecular biology: describing transcription regulatory networks at different scales. *Philosophical Transactions of the Royal Society B:Biological Sciences* 361: 483–94.

Schmalhausen, I. I. 1949. *Factors of Evolution: The Theory of Stabilising Selection*, trans. I. Dordick. Philadelphia and Toronto: Blakeston.

Schrödinger, Erwin. 1944. *What if Life?* Cambridge University Press.

Sesardic, Neven. 2005. *Making Sense of Heritability*. Cambridge University Press.

Shabalina, Svetlana A., Aleksey Yu Ogurtsov, Vasily A. Kondrashov, and Aleksey S. Kondrashov. 2001. Selective constraint in intergenic regions of human and mouse genomes. *Trends in Genetics* 17: 373–6.

Shannon, Claude E., and Warren Weaver. 1949. *The Mathematical Theory of Communication*. University of Illinois Press.

Shea, Nicholas. 2007a. Consumers need information: supplementing teleosemantics with an input condition. *Philosophy and Phenomenological Research* 75 (2): 404–35.

2007b. Representation in the genome and in other inheritance systems. *Biology and Philosophy* 22: 313–31.

2011a. Developmental systems theory formulated as a claim about inherited information. *Philosophy of Science* 78: 60–82.

2011b. What's transmitted? Inherited information. *Biology and Philosophy* 26 (2): 183–9.

in press. Inherited representations are read in development. *British Journal for the Philosophy of Science*. First published online 29 February 2012. At http://bjps. oxfordjournals.org/content/early/2012/02/29/bjps.axr050.full.pdf+html.

Shin, Chanseok, and James L. Manley. 2004. Cell signalling and the control of pre-mRNA splicing. *Nature Reviews Molecular Cell Biology* 5: 727–38.

Simpson, George G. 1953. The Baldwin effect. *Evolution* 7: 110–17.

Sloan, Phillip R., and Brandon Fogel (eds.). 2012. *Creating a Physical Biology: The Three-Man Paper and Early Molecular Biology*. University of Chicago Press.

Smith, Christopher W. J., and Juan Valcárcel. 2000. Alternative pre-mRNA splicing: the logic of combinatorial control. *Trends in Biochemical Sciences* 25: 381–8.

Snyder, Michael, and Mark Gerstein. 2003. Defining genes in the genomics era. *Science* 300 (5617): 258–60.

Sober, Elliott. 1984. *The Nature of Selection: Evolutionary Theory in Philosophical Focus*. Cambridge, MA: MIT Press.

1999. The multiple realizability argument against reductionism. *Philosophy of Science* 66 (4): 542–64.

Stamm, Stefan 2002. Signals and their transduction pathways regulating alternative splicing: a new dimension of the human genome. *Human Molecular Genetics* 11 (20): 2409–16.

Stanford, P. Kyle, and Philip Kitcher. 2000. Refining the causal theory of reference for natural kind terms. *Philosophical Studies* 97: 99–129.

Star, Susan Leigh, and James R. Griesemer. 1998. Institutional ecology, 'translation,' and boundary objects: amateurs and professionals in Berkeley's Museum of Vertebrate Zoology, 1907–39 (abridged). In *The Science Studies Reader*, ed. M. Biagioli. New York: Routledge, 505–24.

Stein, Lincoln D. 2001. Genome annotation: from sequence to biology. *Nature Reviews Genetics* 2: 493–503.

Sterelny, Kim, Michael Dickison, and Kelly Smith. 1996. The extended replicator. *Biology and Philosophy* 11 (3): 377–403.

Sterelny, Kim, and Paul E. Griffiths. 1999. *Sex and Death: An Introduction to the Philosophy of Biology*. University of Chicago Press.

Stern, David, and Virginie Orgogozo. 2008. The loci of evolution: how predictable is genetic evolution? *Evolution* 62 (9): 2155–77.

Stotz, Karola. 2006a. With genes like that, who needs an environment? Postgenomics' argument for the ontogeny of information. *Philosophy of Science* 73 (5): 905–17.

2006b. Molecular epigenesis: distributed specificity as a break in the Central Dogma. *History and Philosophy of the Life Sciences* 28 (4): 527–44.

2008. The ingredients for a postgenomic synthesis of nature and nurture. *Philosophical Psychology* 21 (3): 359–81.

2010. Human nature and cognitive-developmental niche construction. *Phenomenology and the Cognitive Sciences* 9 (4): 483–501.

Stotz, Karola, and Colin Allen (eds.). 2008. *Philosophical Psychology* 21 (3: special issue, Reconciling Nature and Nurture in the Study of Behavior).

Stotz, Karola, Adam Bostanci, and Paul E. Griffiths. 2006. Tracking the shift to 'post-genomics'. *Community Genetics* 9 (3): 190–6.

Sturm, Nancy R., and David A. Campbell. 1999. The role of intron structures in trans-splicing and Cap 4 formation for the Leishmania spliced leader RNA. *Journal of Biological Chemistry* 274 (27): 19361–7.

Sultan, Sonia E. 2007. Development in context: the timely emergence of eco-devo. *Trends in Ecology and Evolution* 22 (11): 575–82.

Suomi, Stephen J. 2004. How gene–environment interactions can influence emotional development in rhesus monkeys. In *Nature and Nurture: The Complex Interplay of Genetic and Environmental Influences on Human Behavior and Development*, ed. C. Garcia-Coll, E. L. Bearer, and R. M. Lerner. Mahwah, NJ: Lawrence Erlbaum Associates, 35–51.

Suppes, Patrick. 1960. A comparison of the meaning and uses of models in mathematics and the empirical sciences. *Synthese* 12: 287–301.

Suzuki, Y., and H. F. Nijhout. 2006. Evolution of a polyphenism by genetic accommodation. *Science* 311 (5761): 650–2.

Tabery, James G. 2007. Biometric and developmental gene–environment interactions: looking back, moving forward. *Development and Psychopathology* 19: 971–6.

2008. R. A. Fisher, Lancelot Hogben, and the origin(s) of genotype–environment interaction. *Journal of the History of Biology* 41 (4): 717–61.

2009. Difference mechanisms: explaining variation with mechanisms. *Biology and Philosophy* 21 (5): 645–64.

Tabery, James G., and Paul E. Griffiths. 2010. Historical and philosophical perspectives on behavioral genetics and developmental science. In *Handbook of Developmental Science, Behavior and Genetics*, ed. K. E. Hood, C. T. Halpern, G. Greenberg and R. M. Lerner. Oxford: Wiley-Blackwell, 41–60.

Tasic, Bosiljka, Christoph E. Nabholz, Kristin K. Baldwin, Youngwook Kim, Erroll H. Rueckert, Scott A. Ribick, Paula Cramer, Qiang Wu, Richard Axel, and Tom Maniatis. 2002. Promoter choice determines splice site selection in protocadherin alpha and -gamma pre-mRNA splicing. *Molecular Cell* 10 (1): 21–33.

Thompson, Paul. 1989. *The Structure of Biological Theories*. Albany: State University of New York Press.

Thornburg, B. G., V. Gotea, and W. Maka lowski. 2006. Transposable elements as a significant source of transcription regulating signals. *Gene* 365: 104–10.

Tilgner, H., D. G. Knowles, R. Johnson, C. A. Davis, S. Chakrabortty, S. Djebali, J. Curado, M. Snyder, T. R. Gingeras, and R. Guigó. 2012. Deep sequencing of subcellular RNA fractions shows splicing to be predominantly co-transcriptional in the human genome but inefficient for lncRNAs. *Genome Research* 22 (9): 1616–25.

Tinbergen, Nikolaas. 1963. On the aims and methods of ethology. *Zeitschrift für Tierpsychologie* 20: 410–33.

Turkheimer, Eric. 2000. Three laws of behavior genetics and what they mean. *Current Directions in Psychological Science* 9 (5): 160–4.

Turney, Jon. 2005. The sociable gene. *EMBO Reports* 6 (9): 809–10.

Uller, Tobias. 2008. Developmental plasticity and the evolution of parental effects. *Trends in Ecology and Evolution* 23 (8): 432–8.

van Fraassen, Bas. 1980. *The Scientific Image*. Oxford University Press.

Venter, J. C., M. D. Adams, E. W. Myers, P. W. Li, R. J. Mural, G. G. Sutton, H. O. Smith, M. Yandell, C. A. Evans, R. A. Holt, et al. 2001. The sequence of the human genome. *Science* 291: 1304–51.

Visscher, Peter M. 2008. Sizing up human height variation. *Nature Genetics* 40 (5): 489–90.

Waddington, Conrad H. 1940. *Organisers and Genes*. Cambridge University Press.

  1952. The evolution of developmental systems. In *Twenty-eighth Meeting of the Australian and New Zealand Association for the Advancement of Science*, ed. D. A. Herbert. Brisbane, Australia: A. H. Tucker, Government Printer Brisbane.

  1953a. Genetic assimilation of an acquired character. *Evolution* 7: 118–26.

  1953b. The 'Baldwin effect', 'genetic assimilation' and 'homeostasis'. *Evolution* 7 (4): 386–7.

  1957. *The Strategy of the Genes: A Discussion of Some Aspects of Theoretical Biology*. London: Ruskin House/George Allen & Unwin.

Wade, Michael J. 1998. The evolutionary genetics of maternal effects. In *Maternal Effects as Adaptations*, ed. T. A. Mousseau and C. W. Fox. Oxford University Press, 5–21.

Wagner, Günter P. 1994. Homology and the mechanisms of development. In *Homology: The Hierarchical Basis of Comparative Biology*, ed. B. K. Hall. New York: Academic Press, 273–99.

  1995. The biological role of homologues: a building block hypothesis. *Neues Jahrbuch für Geologie und Paläontologie, Abhandlungen* 195: 279–88.

1996. Homologues, natural kinds and the evolution of modularity. *American Zoologist* 36: 36–43.

Wahlsten, Douglas. 1990. Insensitivity of the analysis of variance to heredity-environment interaction. *Behavioral and Brain Sciences* 13: 109–61.

Wahlsten, Douglas, and Gilbert Gottlieb. 1997. The invalid separation of effects of nature and nurture: lessons from animal experimentation. In *Intelligence, Heredity and Environment*, ed. R. J. Sternberg and E. Grigorenko. Cambridge University Press, 163–92.

Walters, Ryan D., Jennifer F. Kugel, and James A. Goodrich. 2009. InvAluable junk: the cellular impact and function of Alu and B2 RNAs. *UBMB Life* 61 (8): 831–7.

Waters, C. Kenneth. 1990. Why the antireductionist consensus won't survive the case of classical Mendelian genetics. In *Proceedings of the Biennial Meeting of the Philosophy of Science Association. Vol. 1: Contributed Papers*, ed. A. Fine, M. Forbes, and L. Wessells. Philosophy of Science Association, 125–39.

1994. Genes made molecular. *Philosophy of Science* 61: 163–85.

2000. Molecules made biological. *Revue Internationale de Philosophie* 4 (214): 539–64.

2004. What was classical genetics? *Studies in History and Philosophy of Science* 35 (4): 783–809.

2006. A pluralist interpretation of gene-centered biology. In *Scientific Pluralism, Vol. 19*, ed. S. Kellert, H. E. Longino and C. K. Waters. Minneapolis: University of Minnesota Press, 190–214.

2007. Causes that make a difference. *Journal of Philosophy* 104 (11): 551–79.

Watson, James D., and Francis H. C. Crick. 1953a. Molecular structure of nucleic acids: a structure for deoxyribose nucleic acid. *Nature* 171: 737.

1953b. Genetical implications of the structure of deoxyribose nucleic acid. *Nature* 171: 964–7.

Weaver, Ian C. G., Nadia Cervoni, Frances A. Champagne, Ana C. D'Alessio, Shakti Sharma, Jonathan R. Seckl, Sergiy Dymov, Moshe Szyf, and Michael J. Meaney. 2004. Epigenetic programming by maternal behavior. *Nature Neuroscience* 7 (8): 847–54.

Weber, Bruce H., and David J. Depew. 1996. Natural selection and self-organisation. *Biology and Philosophy* 11: 33–65.

(eds.). 2003. *Learning, Meaning and Emergence: Possible Baldwinian Mechanisms in the Co-evolution of Language and Mind*. Cambridge, MA: MIT Press.

Weber, Marcel. 2005. *Philosophy of Experimental Biology*. Cambridge University Press.

2006. The Central Dogma as a thesis of causal specificity. *History and Philosophy of the Life Sciences* 28 (4): 595–609.

Weiss, Kenneth M., and Anne V. Buchanan. 2009. The cooperative genome: organisms as social contracts. *International Journal of Developmental Biology* 53: 753–63.

West, Meredith J., and Andrew P. King. 1987. Settling nature and nurture into an ontogenetic niche. *Developmental Psychobiology* 20 (5): 549–62.

1988. Female visual displays affect the development of male song in the cowbird. *Nature* 334: 244–6.

2008. Deconstructing innate illusions: reflections on nature–nurture–niche from an unlikely source. *Philosophical Psychology* 21 (3): 383–95.

West, Meredith J., Andrew P. King, and Anne A. Arberg. 1988. The inheritance of niches: the role of ecological legacies in ontogeny. In *Handbook of Behavioral Neurobiology*. Vol. 9: *Developmental Psychobiology and Behavioral Ecology*, ed. E. M. Blass. New York: Plenum Press, 41–62.

West, Meredith J., Andrew P. King, and David J. White. 2003. The case for developmental ecology. *Animal Behaviour* 66: 617–22.

West, Meredith J., Andrew P. King, David J. White, Julie Gros-Louis, and Grace Freed-Brown. 2006. The development of local song preferences in female cowbirds (*Molothrus ater*): flock living stimulates learning. *Ethology* 112: 1095–107.

West-Eberhard, Mary Jane. 1998. Evolution in the light of developmental and cell biology, and vice versa. *PNAS* 95: 8417–19.

2003. *Developmental Plasticity and Evolution*. Oxford University Press.

2005a. Developmental plasticity and the origin of species differences. *PNAS* 102 (Suppl. 1): 6543–9.

2005b. Phenotypic accommodation: adaptive innovation due to developmental plasticity. *Journal of Experimental Zoology, Part B: Molecular and Developmental Evolution* 304: 610–18.

Wilkins, Adam. 2011. Epigenetic inheritance: where does the field stand today? What do we still need to know? In *Transformations of Lamarckism: From Subtle Fluids to Molecular Biology*, ed. S. B. Gissis and E. Jablonka. Cambridge, MA: MIT Press, 389–93.

Williams, George C. 1966. *Adaptation and Natural Selection*. Princeton University Press.

1992. *Natural Selection: Domains, Levels and Challenges*. New York: Oxford University Press.

Wimsatt, William C. 1986a. Forms of aggregativity. In *Human Nature and Natural Knowledge*, ed. M. G. Grene, A. Donaghan, A. N. Perovich, and M. V. Wedin. Dordrecht: Reidel, 259–91.

1986b. Developmental constraints, generative entrenchment and the innate-aquired distinction. In *Integrating Scientific Disciplines*, ed. W. Bechtel. Dordrecht: Martinus Nijhoff.

Winther, Rasmus G. 2006. Parts and theories in compositional biology. *Biology and Philosophy* 21 (4): 471–99.

Wissinger, B., W. Schuster, and A. Brennicke. 1991. Trans splicing in Oenothera Mitochondria: nad1 mRNAs are edited in exon and trans-splicing group II intron sequences. *Cell* 65 (3): 473–82.

Wolkenhauer, Olaf. 2001. Systems biology: the reincarnation of systems theory applied in biology? *Briefings in Bioinformatics* 2 (3): 258–70.

Woodward, James. 2003. *Making Things Happen: A Theory of Causal Explanation.* New York: Oxford University Press.

2010. Causation in biology: stability, specificity, and the choice of levels of explanation. *Biology and Philosophy* 25 (3): 287–318.

Yang, Jian, Beben Benyamin, Brian P. McEvoy, Scott Gordon, Anjali K. Henders, Dale R. Nyholt, Pamela A. Madden, Andrew C. Heath, Nicholas G. Martin, Grant W. Montgomery, Michael E. Goddard, and Peter M. Visscher. 2010. Common SNPs explain a large proportion of the heritability for human height. *Nature Genetics* 42 (7): 565–9.

Index

Italics denote figures and tables.

Printed in the United States
by Baker & Taylor Publisher Services